Gender and Agrarian Reforms

Routledge International Studies of Women and Place

Women of the European Union
The Politics of Work and Daily Life
Edited by Maria Dolors Garcia Ramon
and Janice Monk

Who Will Mind the Baby?
Geographies of Childcare and
Working Mothers
Edited by Kim England

Feminist Political Ecology
Global Issues and Local Experience
Edited by Dianne Rocheleau, Barbara
Thomas-Slayter, and Esther Wangari

Women Divided
Gender, Religion and Politics in
Northern Ireland
Rosemary Sales

Women's Lifeworlds
Women's Narratives on Shaping
Their Realities
Edited by Edith Sizoo

Gender, Planning and Human Rights
Edited by Tovi Fenster

Gender, Ethnicity and Place
Women and Identity in Guyana
Linda Peake and D. Alissa Trotz

Gender and Agrarian Reforms

Susie Jacobs

Routledge
Taylor & Francis Group
New York   London

First published 2010
by Routledge
711 Third Avenue, New York, NY 10017

Simultaneously published in the UK
by Routledge
2 Park Square, Milton Park, Abingdon, Oxfordshire OX14 4RN

Routledge is an imprint of the Taylor & Francis Group, an informa business

First issued in paperback 2011

Typeset in Sabon by IBT Global.

Library of Congress Cataloging in Publication Data
Jacobs, Susie M.
    Gender and agrarian reforms / by Susie Jacobs.
    p. cm.—(Routledge international studies of women and place ; 9)
    Includes bibliographical references and index.
  1. Women in rural development—Developing countries. 2. Women in agriculture—Developing countries. 3. Land reform—Developing countries. I. Title.
    HQ1240.5.D44J33 2009
    305.43'63091724—dc22

ISBN13: 978-0-415-37648-8 (hbk)
ISBN13: 978-0-415-80799-9 (pbk)
ISBN13: 978-0-203-86784-6 (ebk)

Contents

Figures

Acknowledgments

I have many people and institutions to thank for the development, writing and production of this book. Firstly, thanks to my colleagues in the Department of Sociology and in Global Studies at Manchester Metropolitan University for interest and encouragement. The Department of Sociology funded attendance at the first World Forum on Agrarian Reform in Valencia, Spain, in December, 2004 and also helped to fund assistance with bibliographic work. I am grateful to the British Academy and to the Manchester Institute of Global and Spatial Transformations (MISST) for funding for participation in the World Congress of Sociology in Durban, July 2006, where I delivered a paper outlining themes in this book. Thanks to Terry Cox for discussions on peasant differentiation and to Lynne Viola and Steven Wegren for guidance on Soviet sources. Rute Caldeira, Ruth Jacobson, Alan Sillitoe and Chris Porter commented on chapters, and I also thank Rute Caldeira for translation of articles and websites from Portuguese. Thanks to Adam for assistance with typing; Graham Trickey read versions of the book and offered many suggestions for editing. I am grateful to Sarah Mosedale for her detailed assistance with referencing and bibliographic matters.

I would like to thank the following for permissions to reproduce illustrations and photographs: the Hoover Institution Archives at Stanford University for permission to use posters for Figures 4.1 and 5.1; Lieven Soete of Katardat.org for permission to reproduce a Soviet-era photograph from his collection for Figure 4.2; the *Movimento dos Trabalhordores Rurais sem Terra* (MST) for permission to reproduce their logo in Figure 7.1; Katinka Ridderbos and the Internal Displacement Monitoring Centre for permission to reproduce a photograph from Zimbabwe for Figure 8.1.

I am grateful to Benjamin Holtzman, my Routledge editor, for his patience while this book was completed. I owe an enormous debt of gratitude to Janet Momsen and Jan Monk, series editors, for their careful reading of chapters, detailed suggestions, and their advice and support throughout this process.

Lastly, thanks to Adam and Graham Trickey for willingness to step over articles and draft chapters over the last three years. Thank you to Graham for endless cups of tea, for intellectual and political discussions, and for unfailing support. I owe you very much indeed.

Introduction

Current events have pushed the topic of agrarian reform onto the world stage. Increases in food prices have sharpened the focus on means of food production and on the potential of small farming. As one article referring to the greater productivity of small farming states, 'these objects of contempt are now our best chance of feeding the world' (Monbiot 2008:25; see also Chapter 1). At the same time, price increases have made access to land vital for millions of small farmers. In West Bengal, peasant farmers and activists have rioted over the loss of their lands to a petrochemical company and to motorcycle factories (Ramesh 2007, 2008). China's economic boom has left millions landless, and large-scale riots and protests over rural poverty and neglect have become commonplace (Watts 2005, 2006). In Brazil, the Landless Movement (MST) has achieved much success in acquiring land for urban and rural landless people through peaceful land invasions and has presented a model of 'alternative societies' in its rural settlements, post land reform. In contrast, state sponsored land invasions in Zimbabwe have presented a 'model' of chaotic, poorly administered, and politically motivated land reform, and one that has helped push the country into mass hunger and crisis. Such events and movements focused attention upon the varying social and economic consequences of land reforms.

The majority of the world's poor are rural people (International Fund for Agricultural Development [IFAD] 2001). Agrarian reform has long been seen as a key to solving problems of rural poverty, and by implication, global poverty (Rigg 2006). Agrarian and land reform programmes have also affected very large numbers of people across Asia, Africa, Latin America and eastern Europe. Landholdings are highly concentrated in most societies—and they are also concentrated in men's hands (Food and Agricultural Organisation of the United Nations [FAO] 2005: 4; Lee-Smith and Trujillo 2006). Land and agrarian reforms potentially provide a prime opportunity to equalise the position of women vis-à-vis men, both in terms of property holdings and more widely. Yet gender has been a topic marginalised both in concrete policies and in discussions of agrarian reform. The 'story' of most agrarian reforms has been one of missed opportunities in terms of gender equity. Gender divisions have in many cases been exacerbated rather than ameliorated. This is ironic, given that the central thrust of agrarian reform has been development and improved livelihoods for the rural poor. This book aims to redress the male bias of most discussions of agrarian reform and to highlight the intrinsically gendered nature of agrarian and land reform processes.

Gender and Agrarian Reforms integrates work on agrarian and land reforms with that on gender relations, and it puts forward an argument regarding gender and agrarian reform policy. Two main types of redistributive agrarian reform exist: collective and individual household. These differ markedly in their impacts on women and on gender relations.

Land reforms directed to individual households often impact negatively on women, particularly wives. This is because the 'household' is considered to be a unified entity, and men as 'household heads' receive land titles or permits. In this process, women often lose rights and autonomy. Agrarian reforms, which have attempted to consolidate land in larger collective units, have affected women differently. Because these have usually been enacted by socialist or 'left'-leaning governments with partial commitment to gender equity, they have paid some attention to gender issues. Policies have largely consisted of basic marriage reform and of bringing women into 'production'. Women sometimes gain property rights or else are able to earn work points as are men. These changes tend to better the status of married women. However, they leave an important basis of women's oppression, their responsibility for domestic labour, largely untouched. The improved position of women within collective agriculture has given rise to much resistance. Many male peasants feel that their authority as household heads is threatened by such policies, as is the related ability to control women's labour and sexuality. This book considers the current trend to land titling and to market-based land reforms and compares these with redistributive agrarian policies.

This is a comparative work in another sense. Its geographical scope is wide, focusing on Latin America, east and southeast Asia, southern Africa, the USSR/Russia, and eastern Europe; the book also discusses other parts of the world more briefly. These cases are set in a historical context, from the era of the Soviet and Mexican reforms of the late 1920s and 1930s and post World War II agrarian reform programmes, as well as more contemporary changes.

BACKGROUND

If agrarian reforms are 'in the news' in the early twenty-first century C.E., they had been previously off the agenda for a long time. The most important cause of this eclipse was the entrenchment of neoliberal policies and associated economic changes especially since the late 1980s. These bypassed state-centric and redistributionist development policies in favour of market-oriented reforms and retrenchment of the role of the state.

Conflicting trends exist with regard to agriculture. On the one hand, the world's population now consists of more urban than rural residents (Adam 2006). Agriculture is of far less importance than previously in the livelihoods of many rural people (Bryceson et al. 2000; Ellis 2000; Francis 2000), and most rural people have had to diversify income sources to survive. Although many remain wholly or partly land based, other income sources, such as agricultural wage labour, other types of wage labour, and informal sector activities, now often form part of livelihood 'portfolios' (Cameron et al. 2006; Rigg 2006). Often these are combined with short- or long-distance migration so that people move physically off the land. Different views exist about whether deagrarianisation will indeed improve living standards of the ex-rural poor (Francis 2000; Jacobs 2002). In any case, agriculture has become less favoured as a means of earning a living (Marcus et al. 1996; Rigg 2006).

Global trends are of land concentration, the spread of agribusiness and land titling. Land titling, put forward especially by the World Bank (Williams 1996; Manji 2006) has come to supplant the older understanding of land reform as indicating redistribution of land to poor producers (see Chapter 1). Thus, the contemporary context of land reforms differs greatly from the postwar 'classic' period (Bernstein 2004). Titling is nearly always a prelude to the development of a land market (García-Frias 2005), and this is likely to undermine the position of small farmers, especially women farmers.

On the other hand, small and medium-scale agriculture remains of importance in many contexts. Land is not only a space for agriculture but also a place of residence, often a basis for community and in some societies is believed to be linked to ancestral spirits. Along other lines, concern exists about the ecological and social consequences of large-scale agriculture and particularly, of agribusiness. Some producers have turned their attention to more sustainable methods of food production, and these usually involve small (or smaller) scale units. In contrast to the general trend of urban migration, neoliberal policies have in some cases meant a *return* to the countryside. This has occurred in Zambia with the closure of the copper mines and in Russia (Burawoy et al. 2000). In Brazil, some unemployed or working class people wish to leave the overcrowded and dangerous urban areas to farm (Caldeira 2006). Thus, the issues of small-scale agriculture and the socio-economic and ecological conditions under which food is produced will not 'go away'. Agrarian reform is one potential solution to these issues.

PARALLEL REALITIES, PARALLEL LITERATURES

Within academic debates on agrarian and land reform, current attention is focussed on disagreements between those espousing 'populist' or peasant-oriented solutions and neo-liberals. Populists advocate land redistribution within small peasant units, whereas neo-liberals advocate variously, market-based reforms, or the efficacy of large agricultural units. These debates follow on from previous ones between populists and Marxists. In most of this literature, despite different political orientations, the issue of gender is either ignored or submerged. A small, although growing, literature on gender, land rights, and land reform does, however, exist.

The two sets of literature, on agrarian and land reform and on gender within this, tend to address different concerns. The general literature tends to concentrate on the politics of land redistribution, peasant mobilisation, and questions of economic efficiency, as well as technical questions concerning cropping and marketing. That on gender and land reform, conversely, concentrates on intrahousehold matters such as the gender division of labour, women's access to land, and power within households, as well as on kinship systems and gender ideologies. The literatures converge to an extent in that they both examine state policies. However, the general literature usually concentrates on matters such as the extent of reform, its political direction, and outcomes in productivity terms. Literature on gender focuses on the unit of redistribution (e.g. household, individual, or collective). A secondary, although important concern, is the fate of women household heads within agrarian reforms.

The marginalisation of gender issues within discussions of agrarian and land reforms is not easily rectified by simply adding on a discussion of gender. Agrarian reforms have significant impacts on household and gender relations, and some of these are unexpected. Land reforms directed to individual families do not normally set out to disadvantage married women. Indeed, their general intents are usually egalitarian and reforming. Nevertheless, they have had perhaps unintended consequences due to a variety of factors, of which the most important is distributing land to the husband as household head (see Chapter 3). Within collectivist reforms, governments did not foresee the reaction from male peasants in collective schemes to women's new status at work and to attempts to socialise domestic work.

In order to understand such processes, gender analyses must be placed at the centre of discussions of agrarian reform. At the same time, it is also the case that gender relations should not be 'lifted out' of the general social, economic and political contexts of life.

> Issues of rights and resource access and control are now acknowledged, but not necessarily in relation to gender, and rarely through the relational, multi-layered lens which . . . gender analysts of land have seen as important. (Leach in Cornwall et al. 2006: 83)

METHODOLOGICAL ISSUES

These considerations lead on to other methodological issues. The geographical and historical coverage of the book has been dependent upon obtaining secondary sources that discuss gender.[1] The literature discussing gender and agrarian reforms is relatively sparse and tends to focus on particular countries. An exception exists for Latin America, given the large body of work of Carmen Diana Deere and Magdalena León de Leal (eg 1982, 1987, 2001: see bibliography). I have concentrated on case studies where literature on gender and agrarian reform is available. This means that some key cases are not discussed in detail due to lack of published information on gender aspects.[2]

As a comparative overview, this book tends to focus on 'macro'-processes. It contextualises the agrarian reform experience in case studies within and analyses existent gender relations. Each experience of agrarian reform differs from others, given that the agricultural background and state policies will differ. Likewise, each configuration of gender relations has some unique features. However, comparisons across countries and regions can yield valuable insights into similarities, as well as illuminating interesting differences. A focus on macro-processes and comparison does not rule out attention to 'micro'-processes of human interaction, with their potential for fluidity and change. When possible, I include details of case studies on gender and land reform which do include discussions of interactions within households, families and communities. Both the wider 'structural' context and the sphere of human action in smaller-scale settings are of great significance. They should not be seen as competing, as they sometimes are within sociological debates. Rather, micro and macro perspectives complement and enrich one another (see Mouzelis 1990, 1991).

BENEFITS OF AGRARIAN REFORM

As some of the cases discussed within this book indicate, there would be many potential benefits stemming from a gender equitable agrarian reform. An equitable agrarian reform would include equal access to land and other resources and inputs for women regardless of marital status. The arguments given echo those of Agarwal's (2003), as well as Deere and León's (2001), division of benefits of land rights into those based on 'efficiency' and 'empowerment'.

A first argument is in terms of justice. The slogan 'land to the tiller' has long been emblematic of agrarian movements. It is well established that in sub-Saharan Africa women are in most cases the main cultivators, responsible for provisioning their households and feeding the husband, children and often other family members. Increasingly, with feminisation of agriculture, this is also true outside Africa. The global trend is for women to take more responsibility in agricultural production, both in households and in response to opportunities in commercial agricultural work (FAO 2005). Women as the main 'tillers', should receive and be able to control land and proceeds from agriculture. Where women are not the primary agriculturalists, their contribution is usually under-recorded and undervalued.

A second argument can be made in terms of efficiency, that is, securing livelihoods and raising production. It is now often acknowledged that rural women's lack of decision-making power often affects agriculture. Their lack of power can act as a disincentive when they farm (Tripp 2004) and almost certainly affects agricultural outcomes. McCusker's study in Limpopo, South Africa (2004), for instance, indicates that women's marginalised status in agriculture in areas of land reform was one important factor accounting for the unproductive use of land. A large percentage of the population being engaged in food production, but lacking entitlements, puts a brake on agricultural development (Lee-Smith and Trujillo 2006).

Land reform often results in increased incomes and improved livelihood security for smallholders. In many cases, this increases food security (El-Ghonemy 1990). Whether people's livelihoods improve with land redistribution in practice will depend upon a number of factors such as soil fertility, adequate state and other support, the existence of credit facilities, and experience in farming.

A third rationale for women's land rights concerns family welfare (Agarwal 2003). A number of studies indicate that women's enterprise and incomes are often more explicitly oriented to food security than are men's. This is especially evident in Africa (Carr 1991) but is also the case elsewhere (Ghimire 2001). Greater food security is likely to occur because of the cultivation of 'women's' [food] crops such as groundnuts and sorghum, as well as due to their greater propensity to reinvest in farm activities (Kidder 1997). This remains true even in circumstances of livelihood diversification. The World Bank (2003) also cites women's access to land and other resources as crucial to children's—particularly girls'—nutritional status and education. Other evidence indicates that food security does not necessarily increase under male direction of agriculture, even where incomes do. Blumberg (1995) points out that it has not been possible to establish a link between rises in men's incomes and improvements in children's nutritional status in Africa, but such improvements do occur when women's (ie mothers') incomes increase.

Fourthly, land titles or land holding might permit women to take advantage of economic opportunities. Women as farmers face a range of constraints including lack of control over labour and lack of access to credit and to other agricultural supports such as extension advice (Davison 1988a; Momsen 2004). Additionally, women often bear a double or triple (Moser 1993) burden of work. Despite these, in some cases women have been able to seize opportunities either as entrepreneurs individually, or in other cases, collectively. Spring (2002) notes great entrepreneurial activity among African women including in farming. In some Sri Lankan villages, women have banded together, have obtained credit, and have become petty commodity producers of cashew nuts, eliminating middlemen (Casinader et al. 1987, cited in Momsen 2004). A Bugandan study found that by the 1990s, female-headed households were more likely than male-headed ones to purchase land (Tripp 2004: 14).

Fifthly, access to land is likely to increase women's social and economic status within households and communities (Agarwal 1994b, 2003; Jacobs 1997, 2002). The way that Wanyeki conceptualises 'land rights' indicates, for instance, that broader decision-making powers are also entailed.

> Land rights are not only rights to access and to control land as a productive resource but also [rights to] information and to decision-making. . . . And [to derive] benefits from land. (Wanyeki (2003: 2)

It has been uncommon for women to control land, but where they do, indications are that they gain at least a measure of security and power.

ORGANISATION OF THE BOOK

The remainder of this chapter gives an overview of the organisation of the book.

The first three chapters outline general issues and orientations. Chapter 1 concerns agrarian and land reforms in general. It discusses definitions, concentrating on reasons for agrarian reforms, and on types of land and agrarian reform. It also picks out several areas of controversy which are of relevance to gender relations and to the general trajectory of reforms such as the issue of productivity of small and large farms.

Chapter 2 deals with conceptual and theoretical issues concerning gender relations among smallholders, which is the main grouping affected by agrarian reform policies. It concentrates on definitional issues. It looks at types of agricultural producers and whether there is a specific peasant economy as argued by Chayanov (1989). Of what significance is the Marxist critique and its emphasis on class differentiation? The chapter also outlines and discusses meanings of phenomena of great relevance for gender analysis: household and 'reproduction'. These are of importance in peasant- or household-based economies. Lastly, it reviews the concepts of patriarchy and gender subordination.

Chapter 3 returns to the topic of agrarian and land reforms. It gives a general overview of reforms based on redistribution to individual households. This is the main type of land reform attempted in most societies, although affecting smaller numbers of people than the very large-scale collectivist reforms. The chapter analyses land reform in sixteen countries in Latin America and the Caribbean, Asia, and Latin America. It finds the results strikingly similar in many respects despite the wide cultural variations that evidently exist. In general, many land reforms have benefited women as well as men by increasing household incomes and food security. However, because land is usually redistributed to men as household heads, women remain seriously disadvantaged. In some cases, women's subordination has deepened with land reform, and their autonomy and decision-making powers have been eroded.

Part II of the book discusses agrarian reforms along collective lines. Agrarian reforms that collectivise land are ones that have followed socialist revolutions. These have advantages for women, in that aspects of production are collectivised, the peasant/household unit is unsettled, and male household heads lose some authority and control over wives and daughters. However, these have suffered from serious flaws, not least the top-down structures and problems of productivity and food security in many cases. The chapters in this part argue that one (although not the only) reason for this was the resistance of male peasants, who usually expected revolutions to yield a 'peasant patriarchy' (Wiergsma 1991) rather than loss of control over female family members.

Chapter 4 discusses the general trajectory of gender and reforms along collective lines. It first discusses the Soviet example and then presents case studies of eastern European and ex-Soviet states that had collectivised agriculture: Uzbekistan, Bulgaria, Hungary, and lastly, Cuba. Chapter 5 examines the complex case of China. China has had the most radical of reform processes. It is by nature unlike other situations, due in part to its size but also to the rapid industrialisation that has taken place in recent decades and which inevitably affects women's as well as men's lives. Chapter 6, the last in the section, deals with the case of Viet Nam, which remains largely an agrarian society. Redistribution of land has been highly egalitarian, thus Viet Nam is a useful 'test' for propositions put forward here. Liberalisation has meant decollectivisation of agriculture in the north, and with this, productivity increased. Liberalisation or *doi moi* has seen the re-emergence of some elements of traditional lineage practices with regard to gender, as well as greater encapsulation of women inside the home. However, each society and case study has unique features: here, women also have traditional roles as traders which allows them some economic autonomy.

Part III takes up two regional case studies of land and agrarian reforms: Latin America and sub-Saharan Africa. Although these differ in significant ways, in both regions the general trend has been towards liberalisation and land titling.

Chapter 7 offers three Latin American examples. Mexico and Nicaragua have both seen liberalisation of former cooperative structures. Brazil has one of the most unequal land distributions in the world but also one of the most active land reform movements globally. The Latin American cases indicate the regional tendency to view only men as 'farmers' and to marginalise rural women's rights. The chapter examines efforts to 'write' women and gender concerns into agricultural policy and to improve the situations of rural women with regard to land rights. Latterly, in many countries these have taken the forms of individual or joint land titling. The chapter raises the question of the extent to which this might benefit poorer women. Chapter 8 discusses African cases, concentrating particularly on

Zimbabwe as well as South Africa. The land reform in Zimbabwe prior to 2000 contrasts sharply with the 'fast-track' reform after that date. The chapter details the trajectory of land reforms there to argue that the current situation must not be allowed to undermine general arguments for agrarian and land reforms. The chapter also makes reference to debates over land titling and communal rights elsewhere in east and southern Africa, briefly examining Tanzania and Mozambique.

The final chapter concludes the book. Agrarian and land reforms have posed many dilemmas for land as well as for feminist activists and policy makers. Issues of gender equity and 'efficiency' or productivity appear to be played off against one another. Since one of the aims of land and agrarian reform is food security, this cannot be forgotten. However, gender equity is also central to the democratic aims of such reforms. This chapter both recapitulates the book's argument and explores some of the conditions for more gender equitable agrarian reforms in contemporary contexts. Central to the latter are the growth of rural women's movements.

Part I

Theoretical Perspectives

1 Debates Over Agrarian Reform

Agrarian and land reforms across the world have differed widely both in intention and implementation. Agrarian reforms have often been associated with social and political revolution, but sometimes programmes have sought amelioration of an existing socio-economic order. In recent years, the term 'reform' has also been applied to market-led economic restructuring, including of land tenure. While greater equity in landholdings is a common objective, even this is debatable. This chapter asks what agrarian reform is. What are the purposes of agrarian and land reforms, and what circumstances provoke or enable them? How do agrarian reforms differ from one another?

I address these questions by reviewing some of the general literature on land and agrarian reforms and provide an overview of several themes. Many of the studies are gender blind or else mention gender relations only in passing. Thus, I do not deal with gendered implications of land and agrarian reforms but highlight some relevant themes which are discussed later in the book in the context of case studies. The first section analyses definitional disputes. The second discusses reasons and rationales for agrarian reform. The third gives an overview of three controversies of relevance to land and agrarian reforms.

The terms 'land reform' and 'agrarian reform' often overlap but are not precisely the same. Agrarian reform is considered to have a wider meaning than land reform. A situation of agrarian reform covers not only a wide redistribution of land but also provision of infrastructure, services, and sometimes, a whole programme of redistributive and democratic reforms. Land reform refers to a narrower redistribution of land, usually to a limited group of beneficiaries. However, in practice, the two terms tend to be used interchangeably.

WHAT IS AGRARIAN REFORM?

The classic definitions of agrarian and land reform belong to the 'moment' of developmental states. Particularly after World War II and decolonisation, it was assumed that the state and state policy could be a 'motor' of development and societal restructuring. Agrarian reforms are one example of such developmentalist policies. The assumption was that the state would provide support services and that redistribution of income and property would provide overall social benefits. The classic examples are in South Korea and Taiwan, but the Chinese state also played a developmental role. One of the earliest agrarian reforms took place in Mexico from the 1930s, and its aim was developmentalist.

In this paradigm, land reform comprises:

i) compulsory takeover of land, usually by the state, from the biggest landowners and with partial compensation; and
ii) farming of that land in such a way as to spread the benefits of the man-land [sic] relationship more widely than before the takeover. . . . Land reform, so defined is an equalising policy, at least in intention. (Lipton 1974)

Land reform entails change in agrarian structure resulting in increased access to land by the rural poor and secured tenure for those who actually work the land (Ghimire 2001: 7). Small cultivators should obtain greater control over the use of land and better terms in their relationships with the rest of society.

Agrarian reform, then, constitutes widespread redistribution of land. It aims to empower poor peasants and to alter the agrarian and class structure of rural society. Some argue that agrarian reform is therefore a revolutionary political concept rather than a reformist one. Solon Barraclough wrote,

It implies changes in power relationships towards greater participation of the rural poor in decision making at all levels and especially in decisions directly affecting their livelihoods. In other words, it has revolutionary implications. (1991: 102)

In practice, however, redistributive land reforms are often much less than revolutionary and take place for a variety of reasons and in a variety of circumstances. These circumstances range from widespread mobilisation to benefit poor peasants, tenants, and the landless to top-down reforms by authoritarian states. Peru's 1968 land reform, for instance, was instituted autonomously by the military government. Other reforms take place due to external influence. For instance, the extensive and successful land reforms in Japan, Taiwan, and South Korea were instituted due to pressure by the USA, and in order to forestall socialist mobilisation by giving smallholders more of a stake in the system. It was felt that more equitable landholdings would provide a basis for a democratic society (Montgomery 1984; Prosterman and Riedinger 1987). In Latin America, the US-backed Alliance for

Progress carried out land reforms with similar intent. The US government saw land reforms with individual family tenure as the 'perfect package', a solution that would raise rural incomes, boost industrialisation processes, and calm peasant unrest (Deere and León 1987; Thiesenhusen 1995).

Attempts have been made to categorise different types of agrarian and land reforms, commonly according to their wider political and economic intents. The nature of the government enacting reforms and the extent of land redistributed are also significant. Thus, revolutionary, conservative, and liberal land reforms may be distinguished (Putzel 1992). These categories are demarcated by policy with regard to several variables, including the form of property rights, transformation or maintenance of state structures, and the process through which agrarian reform is achieved. For instance, a 'revolutionary' reform has often followed political uprisings that change state regimes. These might expropriate a large amount of agricultural land, redistribute it in collectives, and see such agrarian reform as part of a wider process of social change. A 'conservative' reform, conversely, leaves the basic social and political framework intact and usually redistributes less land. Land tends to be purchased by the state and redistributed to a particular group of cultivators for farming on a family or household basis. For instance, in India 1.5 per cent of land had been redistributed to 2 per cent of producers by 1992 in a society with deep and widespread poverty (Sobhan 1993). 'Liberal' agrarian or land reforms are more ambitious than this, seeking better conditions for rural cultivators but without overall social change. Thus, in Mexico during the 1920s and 1930s, a large amount of land, 43 per cent, was distributed to two-thirds of rural workers (Sobhan 1993). However, Mexico's attempt at rural social transformation was limited by prioritising large-scale capitalist farming (see Chapter 7). Land reform is always a matter of degree (Borras 2005) as well as a matter of intent. The great majority of programmes have been incomplete, either redistributing little land or else allowing landlords continued power (Bandyopadhay 1996).

Typologies such as those mentioned may be useful, in my view, as an aid to understanding. However, actual agrarian reforms present complexities and are not always easily categorised (Barraclough 1991). For instance, individual and collective forms of tenure sometimes are intertwined (Zoomers 2000). Categorisations should not be used as 'boxes' which obscure social complexity.

Neoliberal 'Reforms'

The advent of market-based land tenure changes and reforms since the late 1980s and early 1990s has further stretched the meaning of agrarian reform. In general, the question of agrarian reform fell off the political agenda in the 1980s, only to be 'reinstated' as a matter for debate and action later in the decade (FAO 2000; Borras 2003). However, this is in pursuit of an altered agenda in which neoliberal policies have become dominant, in line with International Monetary Fund (IMF) and World Bank policies. The state is seen as a less important actor and usually has command over declining resources after structural adjustment programmes. The term 'reform' itself has often become part of a neoliberal discourse which has to do with dismantling of welfare and other state services and deregulation of labour markets. In agriculture, this implies privatisation of communal and collective land and titling of individual holdings as well as creation of a land market.

Since the 1970s, the World Bank has promoted land reform along with privatisation, and the International Financial Institutions (IFIs) have designed many such programmes. In the 1980s, titling of land was considered to be the appropriate mechanism for privatisation and economic liberalisation (Toulmin and Quan 2001). Reforms to land tenure were advocated alongside setting up reforms of macroeconomic policy, structural adjustment and orientation to production of export crops and goods (Fortin 2005). The World Bank's newer policy document (2003) sees land as a key 'asset' for the rural poor and emphasises 'productive' use of land, 'efficiency', and profitability.

Reform is not simply change. Many changes may ultimately undermine land rights of the poor (see Chapters 7 and 8). The coalition of peasant and small farmer organisations, *Vía Campesina* ['the peasant way'], for example, launched a global campaign for agrarian reform in 1999–2000 in defence of agrarian livelihoods that has attained a degree of prominence (Borras 2008). A number of rural movements today, including Vía Campesina, reject the term 'land reform' as applied to export agricultural models accompanied by privatisation. The World Forum on Agrarian Reform states,

> The agrarian crisis created by the agro-export model under neo-liberalism is bleak indeed. But despite this, peasant, fishers', indigenous peoples' and rural workers' movements . . . are more alive, more organized and more sophisticated than ever, and are actively engaged in resisting the destructive, dominant model. (*Foro Mundial sobre la Reforma Agraria* [FMRA] 2004)

Throughout this book I continue to use the terms 'land reform' or 'agrarian reform' for redistributionist reforms. To avoid confusion, it is preferable to use other terms to describe rearrangements of land tenure and land titling. In concluding this discussion of terminology, it is pertinent to note that definitions of land reform from whatever quarter have generally ignored any notion of gender rights, despite the emphasis on the 'rural poor'. The majority of the extremely poor in the world, at least 75 per cent, live in rural areas (IFAD 2001: 1). Within this category, women figure large. They constitute the majority of rural people and the majority of the poor globally. One important factor is the minimal extent of female property holding, with some estimates as low as 1 per cent globally (Rai 2002: 97).

REASONS FOR LAND REFORM

In theory, there exists wide consensus about the need for agrarian reform to alleviate rural poverty and hunger (Ghimire 2001: 1). Approximately 60 per cent of the world's population still depends on agriculture as a livelihood source (Lipton 2005). Approximately one-third of the world's peoples depend directly upon natural resources, including land and trees, for subsistence (Sachs, J. 2004). Thus, the need for agricultural land has not become redundant, even in a situation of livelihood diversification (Ellis 2000; Jacobs 2002).

This broad agreement in favour of agrarian reform is somewhat compromised by the different aims and scopes of reform in different countries. In particular, an issue is the extent to which the *status quo* might be disturbed by agrarian change that favours the poor. Nevertheless, many reasons for agrarian reform have been put forward. These encompass the social, economic, political, and ecological.

Social/Individual Human Rights

Land reform is often seen as crucial for the civil and human rights of share-croppers, tenant farmers, and agricultural labourers. Possession of a parcel of land, even a small one, can give some basis to resist the demands and encroachments of landlords. Under the former feudal systems of Europe, Japan and colonial Latin America,[1] such domination was enshrined in law and custom. However, the power of the *hacendado* in Latin America and of landowning classes elsewhere often has feudal echoes in its arbitrariness. On remote estates, landowners may be able to exercise quasi-legal as well as economic powers, becoming in effect the rural political authority. Hacendados until recently often chose to exercise *jus primus noctae*, that is, their customary 'right' to have sexual relations with women tenants on their wedding night (Barraclough 1991).

Personal autonomy may thus be one of the most important gains of agrarian reform (Barraclough 1991). Adding gender into evaluation of agrarian and land reforms, however, means that the 'story' becomes more complex (see Chapter 3).

Economic Rationales

The economic case for land reform is crucial. The three most important reasons for land reform at the economic level are:

  i) to raise agricultural productivity;
 ii) to strengthen food security and to lessen poverty for rural households; and
iii) to facilitate industrialisation by 'feeding the cities'.

Redistribution of land is seen as a way to raise agricultural productivity and therefore to lessen poverty. Redistribution ensures land is utilised fully, given that many large agricultural estates contain underutilised land. Additionally, smallholders are considered to have more incentives than *latifundistas* to make productive investments, where these are possible. Land reforms are not always sufficient to guarantee escape from rural poverty (de Janvry et al. 2001). But they do provide sources of income and also insurance against price shocks. A growing literature shows the importance of land reform to physical capital formation and for economic growth (Dekker 2003). Comparative analysis demonstrates that agrarian reform is important in reducing rural poverty and in raising productivity (El-Ghonemy 1990). However, partial implementation of land reform leaves a situation in which large landowners can exercise great influence over land transactions and over policy (El-Ghonemy 1990). For this reason, more comprehensive reforms are often more efficacious.

A second, related, argument for land reform is its potential role in food security. Possession of, or rights to, land would allow peasant households both to cultivate food crops and to sell any cash crops produced instead of having the proceeds skimmed off by a landlord. This point has gender implications, given the widely cited argument that women attend to household food security across many societies (Blumberg 1988; Kidder 1997; Kabeer and Tran 2002).

A further potential benefit of land redistribution is that it helps to broaden the home market through increasing incomes, consumption, and purchasing power. Industrialisation processes are thereby encouraged. This assumes a model of developmentalist industrialisation rather than industry based on export markets. A negative example is the Soviet Union, where agriculture was deliberately 'squeezed' until the 1960s in order to contribute to industrialisation. In other examples, a relatively egalitarian agrarian reform has contributed to subsequent industrialisation. These examples have been concentrated in East Asia: Japan, Taiwan, and South Korea (El-Ghonemy 1990; Bandyopadhay 1996).

Recently, arguments have also been made for economic rights as 'human rights'. These can perhaps be distinguished from individual rights, as economic rights refer in part to group rights. For instance, the International Covenant for Economic, Social and Cultural Rights (ICESCR) establishes a right to livelihood (see Chapter 9). The right to a livelihood and to food security is being claimed by some social movements as a human right to be campaigned for (Face It Act Now [FIAN] 2004), and this may have implications for agrarian reforms.

Political Rationales

Political rationales for land reform differ, not surprisingly, according to political interests and viewpoints. For many, a prime aim of land reform has been to break landlord power. Especially in societies in which agriculture remains of economic importance, landowners may be important government players and power brokers. This is particularly the case at local and regional levels. But despite the power of landlords, governments have sometimes been prepared to concede reform to quell peasant unrest that might lead to wider radicalisation. Or more positively, they have seen land reform as a way of strengthening the rural poor and transforming them into a new class of smallholders with a stake in society.

Movements for land reform have often been associated with the left, but they can come with a range of political motivations and associations. These may include ethnic and racialised mobilisations in which groups seek to reclaim lands lost to colonists, settlers, or to corporate interests (Christodoulou 1990). Relatedly, such claims can be tied to a nationalist agenda wherein land may symbolise collective identity; the US-backed agrarian reforms in Asia, for instance, had nationalist undertones. Increased demands by landowners or rising expectations among peasants can create conditions for militant movements. Thus, in Cuba, aggressive extension of plantations and the eviction of peasants set in motion an increasingly effective peasant resistance. This later aided the urban revolutionary movement. The rise in numbers of absentee landowners meant that the traditional bond between landlords and peasants underwent change, and systems became more transparently exploitative. A breach in traditional relations also played a part in a peasant movement in Ucureña, Bolivia. This grew from a desire to regain past conditions of land tenure (Huizer 2001) but led to a mobilisation that became national. Other movements such as those in the ex-settler societies of southern and eastern Africa are explicitly nationalist and anti-colonialist.

The Environment

Ecological arguments for land reform have come to the fore in recent years. It is argued that peasants with enough fertile land to farm will be less likely to encroach on rainforests, as occurred in Brazil, or to cultivate and to further degrade low-lying, unsuitable land as in Bangladesh (Handelman 2009). Those who are poor and hungry may be prone to harm or destroy their immediate environments in order to survive. Marginal agricultural lands may be overused or overgrazed grasslands. Alternatively, people are often forced to migrate to cities which already have dense populations and rural areas may become depopulated. Land reform programmes can themselves encroach into fragile environments under pressure to distribute land (Dekker 2003). Agrarian reform, if properly organised and administered, however, can be a potential bulwark against increasing environmental degradation driven by poverty. The Landless Movement (*Movimento dos Trabalhadores Rurais sem Terra*, or MST) in Brazil is attempting to foster sustainable agricultural methods in its settlements, for instance. In general, the hope is that agrarian reforms will promote ideas of stewardship towards the environment.

CENTRAL QUESTIONS AND CONTROVERSIES

Certain questions concerning agrarian reform have been subjects of much controversy and debate. I summarise three here, in order to set a frame for discussion of gender issues. As in most of the material cited so far, these debates proceed without reference to gender. The three controversies concern: the optimal size of agricultural units for efficiency, class factors within land reforms, and land tenure and farm organisation within reforms.

Large Units or Small?

Controversy has long raged over 'efficiency' and the optimal scale or size of landholdings. In his writings on the *Agrarian Question* ([1899] 1988), Kautsky first noted that the usual economies of scale found in industry do not happen in agriculture. That is, small farms are usually more productive than larger ones. The central argument is that peasant family farms are able to utilise labour more intensively than are larger and more mechanised ones. Additionally, it is contended that productivity is lower on large farms primarily because they rely on hired labour, which is more expensive and less efficient than unwaged family workers (Binswanger and Elgin 1993: 34). Relatedly, crops might benefit from close attention not required to the same degree within industry. This phenomenon is termed the 'inverse relationship', or IR. The IR has been used to explain the persistence of family farming and of the peasantry into the era of industrial capitalism and beyond (Lipton 1993; Bernstein 2001).

The IR is crucial in terms of the economic rationale for land reform. If small peasant farming is indeed more efficient than large-scale farming, then this constitutes a powerful argument for redistributive land reform to peasant households. This argument also has clear gender implications, in that it is clear that the 'family' labour discussed is in the main *female* and child labour (see Chapter 2). This point is either missed or is ignored within the literature on land reform.

The importance of the IR is such that it merits further exploration. It should be noted, however, that many 'local' factors within agriculture are significant in terms of productivity. For instance, important factors include overall land concentration, which crops are produced, soil type, how crops and farms are managed, and what technologies and machinery are available. Other demographic factors such as size and composition of families also play a part.

The controversy over the IR has been revisited in the pages of the *Journal of Agrarian Studies*. Griffin et al. (2002) forcefully restate the case for redistributive land reform based on the greater productivity of small farming units. Rather than being inefficient because they cannot afford equipment such as tractors, small farms may adopt different techniques of production. This can lead to differences in productivity. Small farms tend to economise on capital, to cultivate land more intensively and to generate more employment per unit of land (Griffin et al. 2002: 286). Land redistribution would thus raise output and rural incomes, bringing about more equal distribution of benefits.

The assumptions of the IR have been critiqued (Byers 2004; Dyer 2004). The IR is seen as holding for *pre*-capitalist agriculture (Dyer 2004: 49) but as breaking down under the capitalist mode of production (Byers 2004). In particular, large capitalist farms have access to new technologies that are able to enhance economies of scale. Additionally, critics hold that the main reason why small farmers intensify labour is poverty and unemployment, not 'labour preference' (see Sen 1981: 209). The poorest intensify labour because their survival may depend upon doing so (Dyer 2004).[2]

In response, Griffin et al. (2004) have replied that the benefits of land reform are not automatic. Land reform works most effectively when it is part of an overall strategy for rural development, or what I have referred to as agrarian reform. In particular, landlord bias in agricultural policy must be removed (Griffin et al. 2002: 316). Rural needs such as improved access to credit, price reforms, and improved physical infrastructure must also be met. They acknowledge that a number of cases exist in which agricultural productivity *fell* post reform. For instance, in eastern Europe post decollectivisation, people cultivated very small plots. This brings up the point of whether a minimum plot size might exist, although clearly this will vary depending on the context. The authors acknowledge that some truth exists in the 'charge' that small farms may be unable to exploit economies of scale. However, this argument should not be overstated. Small farmers often make cooperative arrangements such as borrowing or leasing equipment. Relatedly, some agricultural functions can be combined via activity splitting. For instance, large combines or irrigators can be contracted out to small farms (Lipton 1993). Lastly, small cultivators may be motivated by desperation, but this in itself is not in my view an argument against the IR.

Are There Classes Within the Peasantry?

A second controversy concerns the role of class differentiation within the peasantry. This relates to the political outcomes of agrarian reforms. As seen previously, most redistributionist land reform programmes aim to increase both the incomes and power of poorer peasants and the landless. Such reform aims to reduce the gap between large landowners and the land hungry and thus to have a levelling effect. The wider and more radical the land reform, the more pronounced such an effect will be. However, many arguments in favour of land reform follow a populist perspective in seeing 'the peasantry' or 'the rural poor' as one, undifferentiated grouping (see Chapter 2). Such a perspective ignores the unequal distribution of resources within rural populations, even among the poor. Leaving aside the question of gender inequality, agrarian smallholders have different resource endowments—land, inputs, livestock, and capital.

In Lenin's formulation, the peasantry could not be seen as one class but was subdivided into:

i) 'wealthy' peasants or *kulaks*: peasants who were not, or not yet capitalists but who had above-average holdings and associated inputs, and who hired in labour on a small/medium scale, as well as utilising family labour;

ii) 'middle' peasants who were mainly self-subsistent, utilising family labour and whose production for the market was limited;

iii) 'poor' peasants who did not hold sufficient land and inputs for family subsistence and who had to resort to wage labour on a temporary or permanent basis; and

iv) Relatedly, the agricultural proletariat was not technically a 'peasantry' at all, but landless workers who resided in the countryside and often had kinship and social links to other groups, particularly to poorer peasants (Lenin 1977).

The vision of land reform for many is the creation of an egalitarian rural sector. However, this depends upon enforcement of strict ceilings on the amount of land that can be held per household and the curtailment of attempts to accumulate property. For instance in the Vietnamese redistribution following decollectivisation (see Chapter 6), strict land ceilings have been enforced. These range from two to four hectares and have prevented strong class differentiation (Watts 1998). Unless land ceilings are low and are enforced, redistribution of land may increase differentiation *between* peasants. Those better endowed with fertile land, livestock, agricultural implements, and machinery, access to credit, and the ability to use these effectively are likely to become wealthier. The livelihoods of others may stagnate. They may become poorer or eventually lose their land.

Moreover, 'classic' peasants are not the only rural poor. Other groups also have an immediate interest in land reforms, such as agricultural workers, whether casually or regularly employed, tenants, and sharecroppers (Christodoulou 1990). More privileged rural groupings such as full-time traders and professionals may retain some rights in land. Thus, the interests of different socio-economic groupings may diverge. For instance, should workers on an agricultural estate receive rights to the land they help cultivate, or should land-hungry smallholders on nearby lands receive the redistributed land instead? This is a current issue in Zimbabwe, where farm workers have been marginalised from the current 'fast-track' land reform process (Sachikonye 2003). In Kerala state, India, splits occurred in a widespread land reform during the 1970s (Herring 1990). Reforms in Kerala resolved one class contradiction between landlords and tenants. But another division was exacerbated between newly-landed cultivators and field tenants. Having acquired land, the new smallholders pursued their economic interests and attempted to become 'larger' peasants. In this situation, clearly further redistribution to field labourers was needed.

It is possible to inhibit the growth of class differentiation within redistributive agrarian reforms. Dampening differentiation would entail setting low land ceilings, as noted. It would also entail making legal provisions concerning alienation of land within certain time periods or to non-agricultural producers. These measures in turn imply both political will and state capacity. Perhaps, needless to say, it is not possible to combine such measures with market-based reforms.

These observations concerning social class do not detract from the fact that reducing differentiation between large landowners and the mass of rural producers is an important aim. This is perhaps evident in the fierce resistance to land redistribution on the part of landowners in various parts of the world. Redistributive land reform does weaken the landed aristocracy or agrarian bourgeoisie and does strengthen smaller-scale producers. However, the trend is toward increased differentiation *among* the beneficiaries. In market-based 'reforms', this is a crucial concern as most of the rural poor are likely to lose any land acquired within free markets.

The question of social class is relevant for gender analysis, although rarely discussed. Firstly, women, like men, will be divided into different socio-economic or class groupings and have different interests on this basis. For instance, in Zimbabwean research in the 1980s, I found that after land reform women like men had differentiated into strata among the resettled peasantry (Jacobs 2000b). Secondly, the point that land reform may increase some forms of 'internal' differentiation is of much relevance. If land reforms potentially widen class divisions within the peasantry, might they operate similarly with regard to gender divisions?

Private or Public Property?

The third controversy concerns the contradictions of private property within land reform. This has been a question for Marxists in particular. Classical Marxist theory concentrates on the role of the industrial proletariat in leading the way towards and securing a 'new' socialist society and economic organisation. Therefore, Marxist analysis has not focused on peasants or on agriculture. This sector has been viewed as either a problem or as simply as subordinate to the urban and industrial working class. The concern was how socialist agriculture should be organised in an era when policies alternative to capitalist ones were possible.

This debate in turn is related both to how the peasantry is viewed and to assumptions about scale and productivity. Peasants as a class have been viewed as conservative and as backward, while agrarian reform is seen as a necessary democratic step away from feudalism. Agrarian reform has also been important in securing the peasantry as allies to the proletariat, the 'leading force' of revolutions. However, peasants are also viewed with suspicion, as petty commodity producers and as small property holders. As such, their class interests are in opposition to feudal landlords and to big capital but differ from those of the proletariat. Thus, it did not follow that peasants' interests were seen as fitting in easily with a collective or socialist state. Rosa Luxemburg ([1913] 1973) wrote in *The Accumulation of Capital* that peasants would fiercely defend their newly won property against a socialist state. The petty agricultural sector would potentially oppose socialist organisation. Political doubts were accompanied by economic ones. A common assumption was that since in industry, larger units are the most efficient, this would also be the case in agriculture. These issues were debated at length in the Soviet Union in the 1920s, particularly between Bukharin and Preobrahzhinskii (Lewin 1968), and in the less well-known contributions of Kritsman (Cox 1986).

Different shades of opinion existed about these complex questions. However, a common view among Marxists and other socialists was that agriculture should be collectivised. This would at once solve the perceived political problem of the peasantry and the economic one of 'low' productivity in small peasant agriculture. In practice, the economic results of Stalinist collectivisations in the Soviet Union and many other societies were sometimes disastrous and usually failed to enhance productivity. Top-down state-imposed collectivisations were usually highly unpopular (Lewin 1968; Lipton 1993) and drastically underestimated peasant resistance to collective forms. It is important in this context, however, to raise the question of gender and of women's potential interest in collective forms (see Chapters 4 to 6).

CONCLUSION

From the discussions in this chapter, a number of questions emerge which are relevant to the success or otherwise of land reforms. These include what the purpose of the agrarian reform was seen to be. For instance, was this envisaged as part of a wider social transformation in the countryside? How important were factors such as food security and lessening of differences and inequalities based on class, caste, and ethnicity? What supports, if any, were put in place following the reform? How was the reform carried out, and how complete was it? Such questions tie in with others concerning the 'nature' of the state, whether authoritarian or democratic. Relatedly, how tolerant is the state of social movements, including land movements?

The remainder of this book focuses on gender issues and discusses a number of cases of gender within agrarian reforms. However, gender cannot be viewed apart from the general context of these reforms. Although women and men are affected differentially by many social institutions such as kinship, marriage, and gender ideology, they are also affected by broader social processes. It is thus necessary to discuss the general circumstances of agrarian reforms as well as policies specifically oriented to women. This is not to argue that broader factors must take precedence over gender analysis, or that it is acceptable to view gender as an 'add-on' category. Wide questions of politics, political economy, and cultural changes affect gender relations, and conversely gender should be seen as integral to these processes. Women are often marginalised within agrarian reforms, but they are not marginalised in the same manner or to the same degree in each case.

This chapter has emphasised competing definitions of agrarian and land reforms, particularly between redistributionist and market-based models. Many reasons for agrarian reforms exist, encompassing the economic, political, and ecological spheres as well as that of human rights. The reasons emphasised tend to vary according to the political aim as well as the scope of agrarian reforms: for instance, is equity or efficiency seen as the main goal? The important issue of the optimal size of agricultural units has in practice overlapped with that of how peasants are seen as political actors. Where a benign view is taken, land is usually redistributed to individual household units.

Within these sometimes complex debates, little attention has been paid to the issue of gender divisions. Issues are often discussed as if gender neutral, although in fact programmes have important effects on gender relations and on women's status. Whether an agrarian reform is 'revolutionary', 'liberal', or 'conservative' in its policies will affect all people in terms of the amount of land redistributed. Additionally, women are usually denied the right to hold or to work on this land on the same basis as men. General aims such as increasing production, or creating a more democratic rural society, similarly impact on both sexes. It appears as if women tend to gain more when egalitarian goals are emphasised. 'Revolutionary' or more radical models have been most concerned to target women as actors in their own right. This is partly due to their emphasis on collective forms rather than individual family models. Taking another issue, one might be forgiven for assuming that gender was a taboo subject within discussions of the IR. As has been noted, gender is important to the IR, as those whose labour is intensified are usually female. Agrarian reforms may aim to give small producers more say over their own lives, but this is not possible as long as producers are envisaged as male, and women's work is ignored. Both men's and women's lives are affected through agrarian reforms but not necessarily in the same ways.

Understanding these processes entails knowledge about the general background of agrarian reform as has been explored here. However, it also involves analysis of categories and concepts employed in gender analyses of land rights and agrarian reforms. The next chapter turns to this discussion.

2 Concepts for a Gendered Analysis of Agrarian Reform

The household [in the domestic mode of production] is as such charged with production, with the deployment and use of labor power, with the determination of the economic objective. Its own inner relations, as between husband and wife, parent and child, are the principal relations of production. . . . The built-in etiquette of kinship statuses, the dominance and subordination of domestic life, the reciprocity and cooperation, here make the 'economic' a modality of the intimate. . . . [D]omestic decisions . . . are taken primarily with a view toward domestic contentment. Production is geared to the family's customary requirements. Production is for the benefit of the producers. (Sahlins 1974: 77)

The rigid patriarchy of peasant society was hardly unique to Russia. In Russia, however, the peasant woman was subordinate not just to one father or one husband, but to the entire male community. . . . Women were represented neither in the [peasant] commune nor in the [peasant] courts. . . .

  Thus the peasant woman had no direct access to land, the most vital component of subsistence, nor a role in the conduct of domestic or communal life. She was mute and powerless. . . . Her contribution to family survival was nonetheless great. (Glickman 1992: 55)

Is subordination of women integral to peasant economies? Gender issues have not generally been included in discussions of agrarian reform, whether by social movements or by governments implementing such policies. *One* of the reasons for this is lack of integration of gender issues within theoretical discussions of peasants. Women's work on farms and within homes is fundamental to peasant/smallholder households, but their work is often rendered invisible. Similarly, gender has often been invisible within peasant studies.

  The chapter begins with discussion of peasants, the main subjects of agrarian reforms. It then moves to discuss two concepts integral to study of agricultural smallholders: the 'household' and 'reproduction'. The third section discusses the concept of patriarchy, which attempts to theorise women's subordination across cultures and societies. Lastly, an underlying question concerns whether a 'peasant patriarchy' exists, and this topic is taken up in the final section.

The concepts have been heavily contested. I hope to clarify them but at times am able only to trace lines of disagreement. It is a matter of judgement whether to keep using or to discard a concept or category despite imperfect specification. In most cases, I opt for pragmatic solutions.

THE PEASANTRY?

Agrarian and land reforms usually involve groups of smallholding agriculturalists—or peasants. Most reforms redistribute land to 'household' or family units (see Chapter 1). Land and agrarian reforms may actually *constitute* a peasantry from a landless proletariat. In the Cuban case, for instance, the government redistributed land to landless households as well as forming state farms. More commonly, land is redistributed to poor peasants or sharecroppers, as in India, Brazil, and Nicaragua. Agrarian reform may be along collective lines, so that agricultural labourers remain wage labourers, but with the state as new 'proprietor'. In cases such as the USSR or China, however, most state farm workers had previously been smallholding peasants. Lastly, in a number of cases land had been collectivised but has since been decollectivised (see Chapters 4 to 6). Here, the redistribution of land re-configures a peasant grouping, as in parts of Eastern Europe, Russia, Viet Nam, and China.

The term 'peasant' has been much theorised and debated. I concentrate on discussion of Marxist and populist conceptualisations as these are most relevant to the literature on agrarian reform. However, I make brief reference to other theories. Concepts of peasantries divide broadly into two. The first sees peasants as embodying a particular and separate *type of economy* with associated behaviour. The second focuses on how small agrarian producers are inserted into, and influenced by, the wider capitalist economy. The most influential writer in the first grouping was A. V. Chayanov (1989), writing in the populist school. Others such as Shanin have based their work on Chayanov's theorisation. By comparison, Marxist and neo-Marxist writers have usually critiqued the term 'peasant', refining it or sometimes rejecting it altogether. Most Marxist writers see peasant production units as household economies based on simple commodity production.

Other theoretical trends cross-cut or in some cases stand aside from these two main schools. Neo-classical economists do not fall into either camp as such: they accept the term 'peasant' but focus instead upon household behaviour of micro-farmers. Recently, postmodernist theorists of peasant behaviour have become in vogue, especially in Latin America (Kearney 1996). These concentrate both on formation(s) of identity and on modes of resistance. Previous work within anthropology concentrated on distinguishing peasants from other types of rural cultivators. Robert Redfield's influential work characterised peasants as 'part societies with part cultures' (Redfield 1956). He concentrated upon the distinction between folk [village] culture and that of the dominant 'civilisations' within which peasant cultures were embedded. In Africa, definitional discussions distinguished communal cultivation in 'simple', lineage-based societies from that by peasants in more complex societies (Post 1972).[1] Among peasants, land is usually individually held, kinship and political hierarchies are separated, and market principles dominate. Rent is usually paid to landowners or other powerful groups. Other conceptions emphasise the class nature of societies in which peasants live. Peasants can exist in different types of society—feudal, agrarian/bureaucratic, and capitalist (Wolf 1966).

Chayanov

Chayanov and his colleagues in the Organisation and Production School worked in the Soviet Union during the 1920s. Chayanov held that a peasant mode of production should be accepted alongside those already established in Marxist theory: primitive communist, slave, feudal, and capitalist modes of production. Chayanov's theory was also one of peasant behaviour at the level of the individual family farm. Peasants, or the household head, attempted to balance subsistence needs against the size of the product (the crop) and the 'drudgery' or work intensity necessary to produce sufficient for subsistence (Thorner 1966). Thus, the peasant producer would make an increased effort only if he had reason to believe it would yield a greater output for investment or consumption. The 'labour-consumer balance' prevailed in peasant economies. The intensity and difficulty of work had to be weighed against any increase in output which might result, since the overall aim was not accumulation. Chayanov's theory also rejects the Marxist idea of class differentiation as occurring among the peasantry.[2]

Other conceptions of peasantries have followed aspects of Chayanovian theory. Sahlins (1974) posited the idea of a 'domestic mode of production' based on norms of kinship and reciprocity. James Scott's work (1985) on 'everyday forms of resistance' emphasises village solidarity. Shanin's (1974) characterisation of the 'pure peasant type' draws particularly upon Chayanov and upon anthropological work. The pure peasant type is one in which:

i) the family farm is the basic unit of organisation;
ii) agriculture is the main means of livelihood;
iii) a specific culture relating to the peasant way of life (including conformity, circular perceptions of time) exists; and
iv) peasants are exploited economically and oppressed politically.

In 1990, Shanin revised this definition. He argued that peasant economies consist of a blend of self-employment (family labour); control of own means of production; self-consumption of produce; and multidimensional occupational expertise (1990: 52). He also added the criterion of a specific type of

peasant politics. Peasant political processes are seen as distinctive, characterised by patronage, brokerage, and village factionalism (1990: 53). In this
view, a globally similar peasant social organisation does exist.[3]

Chayanovian conceptions are problematic in several respects. The first is
overgeneralisation on the basis of a European model. Secondly, Chayanov
assumed a basic homogeneity among peasants.[4] In taking the family labour
farm as the basic unit of analysis, he sidestepped the question of wider social
factors such as the influence of the state or the capitalist economy (Ennew
et al. 1977; Patnaik 1979; Harrison 1982). Thirdly, it is doubtful whether
peasants or smallholders can constitute a separate mode of production.
Peasants commonly exist as subordinate segments or modes within a wider
society, dominated by feudalism or capitalism (Banaji 1977).[5] Fourthly,
Chayanovian-influenced theories are problematic in terms of gender analysis. This is unfortunate since the theory is premised on a type of family
farm in which household labour provides the main component. That this
labour is mainly that of women and children is seen as natural and so no
further analysis is attempted.

Despite these criticisms, many of the processes Chayanov highlighted
refuse to go away completely. The kinship basis of economic life, demographic cycles which relate to social differentiation, and the balance
between work and leisure are all features of smallholder households and
economies. In practice, many analyses combine features of Chayanovian
and class analysis (see e.g. Murray 1981).

Marxists and Petty Commodity Production

Whatever the problems of Chayanovian theories, they do posit a clear formulation of 'the peasantry'. In Marxist thought, the category of 'the' peasantry is slippery and ill defined. There are two main Marxist orientations
to the subject of the peasantry and peasant economies. The first deals with
peasants as a *class*, one which is unstable and prone to divide because of the
logic of capitalist economic processes. Lenin analysed the class differentiation of the peasantry positing three strata of peasants, poor, middle, and
wealthy (*kulaks*; see Chapter 1). The discussion then revolves around the
problem of class differentiation. Poor peasants are viewed as quasi-prole-
tarian, and wealthier peasants as proto-capitalists. Relatively self-sufficient
middle peasant households own their own means of production but hire in
little labour and instead rely on 'self-exploitation'.

The second orientation discusses the peasantry with reference to petty
(or simple) commodity production within capitalism. Ennew et al. (1977)
critiqued the concept of peasantry as essentialist. They proposed that
it be replaced by the concept of petty commodity production as more
appropriate for Marxist analysis. This intervention sparked off debates
concerning the term 'peasant', its scope, and applicability.[6] Those party
to the debate agree that wage labour is absent or else sporadic within

petty commodity production. Implicitly or explicitly, they accept that there exists a 'middle' peasantry which poses ambiguities for Marxist analyses. 'Peasant' in this view refers to commodity producers who possess the necessary means of production and who engage in production on the basis of unpaid household labour alone, that is, without wage labour (Gibbon and Neocosmos 1985).

The debates over petty commodity production focused almost exclusively on the relations of petty commodity production with capitalist forces rather than on the internal organisation of the household. Both Gibbon and Neocosmos (1985) and Bernstein (1988a) acknowledge that gender and family relations are important to petty commodity producers. However, these institutions and relationships are explained only in class terms. Men (patriarchs) are analysed as representing the 'place of capital', and women and children that of labour within the peasant household (Gibbon and Neocosmos 1985: 202–3). This overlooks the limitations of class analysis in discussions of women's position.

In practice, virtually all concepts of peasantries find it difficult to analyse issues of gender division and gender subordination. This observation extends to the 'household' economic schools which conceptualise peasant household behaviour (Becker 1976; see Ellis 1988).[7] The concept of petty commodity production does not inevitably posit the idea of a unified household, however, and this means that it is a concept potentially open to gender analysis. Whatmore's (1991) analysis of English family farms, for instance, is able to demonstrate the gendered nature of production even within a near-fully commoditized economy. She also demonstrates the need to dispense with orthodoxies holding that class is the only relevant force.

The omission of gender relations has somewhat different implications for different theories. For Marxists, serious analysis of gender relations among petty commodity producers means that the *a priori* primacy given to class must be queried. Other implications are that exclusively macro-level analyses are incomplete and must intertwine with recognition of human agency. For Chayanovian-derived theory, the implications of gender analysis are most serious: analysis of the 'peasant' unit in gender terms disrupts many of the theory's basic assumptions. This school has been highly influential, and its influence has extended indirectly to the ways that policy makers have viewed peasant households. The failure of analyses of agrarian reform to incorporate gender stems in part back to male-biased concepts (Elson 1995).

SOME FEMINIST CONCEPTS FOR THE ANALYSIS OF PEASANTRIES

I now turn to analysis of concepts central to discussion of small-scale agrarian/peasant economies: the 'household' and 'reproduction of labour'. As these are central to analysis of peasantries, they are also relevant for discussion of gender and agrarian reform.

The Household

Households are crucial to the study of smallholder economies: the household is a site of agricultural production as well as of domestic life. Household labour and household decision-making are crucial to the functioning of peasant households. This feature also distinguishes peasants/smallholders from many other types of producers. For instance, both slaves and proletarianised workers labour in wider economies as well as within households. Households are of particular importance to women because they are associated with the domestic. In many societies, women live out large portions of their lives inside physical homes, and their mobility outside may be severely restricted. Households are sites where gender relations are enacted, reinforced, and altered. As Henrietta Moore notes,

> Households are important to feminist analyses because they organise a large part of women's domestic and reproductive labour. As a result, both the composition and organisation of households have a direct impact on women's lives and in particular, on their ability to gain access to resources, to labour and to income. (1988: 55)

The household is sometimes considered to be coincident with 'the family', particularly for neo-classical theorists. Families are usually seen in anthropological and sociological writing to be constituted through a combination of marriage, kinship relations, and economic activity. More usually, the 'household' is considered to be a somewhat broader term, constituted similarly to the family. However, less stress is placed on links through marriage and kinship and more on residence, production, and commensality. Two issues are of relevance: (i) that of household forms and boundaries and (ii) that of internal processes within households.

Anthropologists have been active in excavating variations in household structure and composition. Forms of marriage, forms of residence, the activities which take place within household/domestic domains, the extent of separation from the extra-domestic sphere, all of these vary. Households among peasants may be nuclear or extended. Marriage forms can be polygynous, monogamous, and (occasionally) polyandrous. In practice, however, *most* agrarian households are small, containing less than six members, and tend to be fairly simple in organisation (White 1980: 16). Smallholder households are usually monogamous, at least in the formal sense. Even where polygyny is permitted, most men cannot afford to marry more than one wife.

Kinship systems are important in framing household relations. Most of the cases discussed in this book are either patrilineal or bilateral: that is, calculating kinship on both sides. Patrilineal systems predominate in much of Africa, the Middle East, and east and southeast Asia. However, 'pockets' of matrilineality have existed, particularly in west and central Africa and south and southeast Asia. Although still male dominated, matrilineal systems typically give women stronger rights, including rights to land (Agarwal 1994b; Stivens et al. 1994). Patrilineal systems vary: in some, women retain attachment to their father's lineage and therefore to their relatives. In extreme cases such as Han China, women are considered to be completely absorbed into the husband's lineage (see Chapter 5). Patrilineality is often accompanied by patrilocality, which is for the most part disadvantageous to in-marrying wives. However, in some systems the preferred form of marriage is between cross-cousins: in this case the in-marrying wife is actually a relative, and this may give her more influence. Bilateral kinship predominates in Europe and in most of Latin America and the Caribbean. It is usually associated with neo-local or virilocal residence and either lack of lineage structure or weak lineages. These factors usually advantage wives, and it has been argued that in European peasant households wives are influential in decision-making (Rogers 1980).

Whatever the form of marriage and household, most societies have 'family ideologies' (Barrett and MacIntosh 1982) or idealised models of households and family relations to which many aspire (Wilk 1984). For instance, in pre-revolutionary Viet Nam and China, the ideal family was seen as a five-generation patrilineal and patrilocal family, along Confucian lines. In practice, very few families and almost none outside the elite could achieve this. When a family had no son, sometimes a son-in-law was imported and the daughter remained in her father's home. Although having advantages for wives, this was, and is, considered to lower the husband's status (Bossen 2002). In contemporary western Europe and parts of eastern Europe, the model family is monogamous, heterosexual, and nuclear in form with two children—ideally a boy and a girl. Such family models can have profound effects. Those households which adhere most closely to the ideal model are usually accorded higher status and people in other family types may be considered inferior. In many societies, households headed by women are not considered proper families. Female-headed households were, at least until recently, often neglected when policies were formulated.

Discussion of household boundaries has also concentrated on variability. In some cultures, the concept is absent (Wilk 1984). In other cases 'the' household may be difficult to define. African households, in particular, typically have permeable boundaries. In other situations, however, definitions of boundaries are clear. In China since the 1960s, households have been identified through the household registration system. Each person is definitively recorded as belonging to a household in a specified locality and of a specified type, rural or urban. Aside from its other functions, this system precisely registers the boundaries of each household (Judd 1994).

Within agrarian reforms, the state or other bodies may become involved in regulation of various household matters, including household boundaries. For instance, it is common to allow only one wife to register as a family beneficiary, along with only some children even where polygyny is widespread (see Chapter 3). Policing of behaviour within the household boundary may also take place. This may extend to control over sexual behaviour or over other behaviour such as male violence.

Feminists have critiqued most theorisations of peasantries for assumptions that households are unified, as noted previously. Households can be seen as unified because they are viewed as headed by autocrats taking decisions on behalf of the family, as in neo-classical theory (see Ellis 1988). Alternatively, households can be conceptualised as unified if they are assumed to be governed by 'moral economy' (Sahlins 1974; Scott 1985). The idea of moral economy posits a more benign view of household processes. Peasant families are assumed to unselfishly care for one another (Folbre 1986b: 7), and household goods, food, and property are pooled. In this view, kinship relations operate through norms of generalised reciprocity. Household relations are egalitarian, without strong status or power differentials. A related assumption is that separate but equal spheres exist for males and females. That is, the existence of a gender division of labour is seen to give women behind the scenes household power (Rogers 1980). Sen's idea (1990) of 'cooperative conflicts' offers a nuanced view of households as sites both of unity and dissension. In most feminist thought, however, the household is seen as *divided* through differential power of men and women. The household is often seen as the privileged site of oppression and exploitation of women, although it is not the only site. This discussion is continued in the next section, considering the idea of patriarchy.

Reproduction and Reproduction of Labour

Attention to household processes has also been associated with analyses of 'reproduction'. The concept of 'reproduction' is important in the context of discussion of agrarian reform in several ways. Firstly, peasant/smallholder units combine processes of production of crops and crafts for sale or exchange with household reproduction. Secondly, socialist governments have tried to collectivise production, but it has been very rare for questions of reproduction to be addressed. Instead, domestic work is usually naturalised and simply considered to be part of women's existence. However, a few attempts have been made to socialise domestic labour (see Chapters 5 and 6). These have been met with strong resistance on the part of male peasants and sometimes from groups of women as well. Thirdly, in agrarian reforms which distribute land to individual families/households, women's work may increase due to the lack of separation between productive and reproductive labour. Discussion of reproduction inevitably requires some discussion of production.

Edholm et al. (1977) pioneered critiques of the term 'reproduction', pointing out that the concept had three main meanings:

    i) biological reproduction (of children);
    ii) reproduction of labour; and
    iii) general social reproduction, or ensuring the continuity of social and
       economic arrangements.

Their three-fold division is generally accepted. The first two senses of reproduction are particularly associated with women and with the domestic/household domain, and women carry out much of the work involved in each.

The meanings of 'reproduction' in the sense of biological reproduction and of actions and processes ensuring social continuity are not contentious. The idea of general social reproduction is often used descriptively in the social sciences. The second meaning of 'reproduction', however, is more problematic. Particularly in analysis of agricultural smallholders, many activities are not commoditised. Thus, it may be unclear what is productive and reproductive.[8]

In most societies, women[9] are responsible for childcare, early socialisation, cooking, cleaning, and many repetitive chores, the exact nature of which vary. The assignment of such work to women is an economic expression of the inequality of the marriage contract, and the nature of this contract is especially important in reproducing women's social position (Whitehead 1981). Gender typing of work seems to be most rigid in tasks crucial to the perpetuation of relations of marriage, filiation, and procreation (Mackintosh 1981: 11).

In advanced capitalism, such work appears as 'housework', separate from the productive sphere of waged work. In small-scale agrarian households, many activities are not commoditised and remain use-values. In such economic systems, 'domestic' labour may be characteristic of all household members' labour across a variety of tasks. Moreover, the same task can either produce use- or exchange-values depending upon the context and regardless of gender. For instance, cooking food in the home is 'reproductive' while cooking food for sales in the street is 'productive' (Redclift and Mignione 1985: 97). Likewise, raising food for family consumption is reproductive while cash cropping of the same items is productive. The term 'reproduction' in practice tends to be used in a sleight-of-hand manner to identify *women* and their work (Harris 1981). In literature on gender and land reform, the term 'reproductive labour' is often used, generally to signify domestic labour, childcare, and agricultural activities such as cultivation of vegetables for family consumption. I [also] reproduce the term 'reproduction' here to describe such work, while remaining uneasy about the category.

This usage can also perpetuate the idea that reproductive work is somehow 'not work': it does not produce value, is less valued socially, and is often invisible. Some suggest jettisoning the distinction between production and reproduction (Kusterer 1990; Whatmore 1991)—unless one refers to the labour of childbirth. Miriam Glucksmann (2005) provides a way forward, in her conception of the 'total social organisation of labour' which demonstrates the interconnectedness of work undertaken in different socio-economic relations and spaces. Thus, the total social organisation of labour argues for conceptualisations of work to include unremunerated as well as paid work and which may be conducted in terms of social or kinship relations (Glucksmann 2005: 23).

Patriarchy?

The final concept discussed here is that of patriarchy. This concept broadens the discussion to the social realm in general. Gender relations are not confined to the household, even where women are. Households are always part of wider communities, economies, and societies. The concept of patriarchy has a long history but has been elaborated particularly within feminist writing since the 1970s. Like other concepts discussed here, different meanings have been attached to it.

Patriarchy means, literally, 'the rule of the father'. It has been used since the seventeenth century to denote the rights of a ruler, representing the ultimate power of the king as a father over his 'family'. Maine ([1861] 2001) used the term in his study *Ancient Law* and described the absolute supremacy of the eldest male parent of the household, his powers extending to life and death. By the nineteenth century, 'patriarchy' came to refer specifically to household relations. Max Weber, writing at the end of the nineteenth and early twentieth centuries, was concerned with systems of organisational domination. For him, patriarchy occurred when the father had unconstrained power within a household (Waters 1989). For Engels (1972), patriarchy was conceptualised as specific to a period of history *prior* to the development of the state; when the state appears, it appropriates much of the authority previously accruing to household heads. These views also acknowledged the subordination of women in particular, as well as of junior men. These ideas, then, specify a form of patriarchy in which the rule of the father is supreme.

In the 'second wave' of writing about patriarchy, a much broader and less household-based view was presented. Millett in *Sexual Politics* (1970) takes power relations within families as the starting point, seeing patriarchy as an institution through which all women are controlled by all men. This power is all pervasive, therefore appearing natural and incapable of change. Patriarchy overrides class and racial divisions and is primarily maintained by socialisation, as well as by the use of force (Millett 1970). Others agree that patriarchy does not derive from capitalism, or indeed from any other system, but is *the* basic organisational system. Male supremacy derives from appropriation of women's bodies, from compulsory heterosexuality and ultimately, from male violence (Brownmiller

1975; Rich 1986). Thus, patriarchy appears as a more or less universal system with little social variation.

This use of patriarchy in this sense sparked off a wide debate concerning the term's adequacy. Marxist and socialist feminists criticised the lack of attention to class divisions (see Rubin 1975; Harris 1981; Rowbotham 1982). For instance, lack of land reforms disadvantages all the rural poor, including women. It affects women in terms of their *class* positioning (e.g. as landless workers, sharecroppers or unemployed people). Some Marxist feminists posited a 'dual systems theory', giving capitalism and patriarchy equal weight (Eisenstein 1979; Hartmann 1979; Mies 1986) and also affirming that individual men of subordinated classes, not only 'capitalists', benefit from women's subordination.

Another criticism concerns the overgenerality of patriarchy and lack of acknowledgement of variations across cultures. Such arguments are of much relevance to this book, which attempts a comparative analysis of gender relations. From the 1980s, black and ethnic minority women critiqued the idea as ethnocentric. It was argued that for many women, state and more general social racism might be a greater source of oppression than male household domination (hooks 1982; Amos and Parmar 1984; Ramazanoglu 1989; Mirza 1997). This analysis parallels those of socialist feminists in objecting to one axis of oppression (gender) being emphasised, while others such as ethnicity or class are downplayed across cultures. Many of the poorest people in Latin America, Asia, and elsewhere are, for example, indigenous peoples, now made ethnic minorities. Often, they have been dispossessed of their land through colonial and other incursions. Their struggles for agrarian reform today involve regaining land, not only to cultivate but also as the basis for their societies. Thus, their movements tend to centre on ethnic/indigenous identities (Deere and León 2000).[10] More recently, discussions of intersectionality have addressed the problem of which division(s) should be prioritised (e.g. Anthias and Yuval-Davis 1992; Yuval-Davis 2006), arguing that divisions should be considered simultaneously and that the importance of any particular form of oppression may vary according to context. The term 'women's subordination' thus tends to be associated with critiques of patriarchy (see Whitehead 2006).

Another theoretical criticism concerns the failure to specify the driving forces of patriarchy (e.g. Acker 1989; Pollert 1996). Attempts to answer this criticism argue that patriarchy is a system in the same way that capitalism is a system of production. Mitchell in *Psychoanalysis and Feminism* (1975) first specified structures of patriarchy, following an Althusserian model. The structures were production, biological reproduction, sexuality, and socialisation. Production was seen as dominated by capitalism while the latter three factors were dominated by patriarchy through the operation of the unconscious.

Walby (1990) outlines six structures:[11] patriarchal relations within paid work, within the state, male violence, sexuality, cultural institutions (e.g. religion, education, and the media), and the patriarchal mode of production. The latter concerns household relations and especially the performance of domestic labour conceived as near-exclusively female. Her concept of a domestic mode of production within patriarchy relies on the idea of a 'domestic mode'. This posited that men and women, especially husbands and wives, are in different and always opposed classes (Delphy 1977).

More recently, feminist work has emphasised agency, diversity, and contingency, in accord with postmodern currents (Butler 1990; see Segal 1999). Tinsman (2002), in analysing Chilean agrarian reform, writes that patriarchy is not a 'master grid' but 'a variety of arrangements derived from broad principles legitimating men's authority over women. . . . These arrangements undergo constant negotiation and change' (12–13). Another reaction has been to avoid the use of the term 'patriarchy', as in the term 'female subordination'. Connell (1987), for instance, uses the term 'gender order'. Gender orders are constituted in three gender practices:

- power: ways that men and women control one another;
- labour; and
- 'cathexsis': ways in which attachments between people are formed.

Connell also employs the useful term 'gender regime' for particular gendered situations.

The concept patriarchy does not adequately characterise every situation of male domination. As alternatives, 'gender subordination' and 'gender regime' capture the systematic nature of male bias, along with openness to variation and human action. For this reason, I often employ these concepts. However, I feel that 'patriarchy' should be used when its classical sense applies. Some of the situations analysed here are ones in which women as a group are extremely subordinate. In these cases, I use patriarchy. Where other authors use patriarchy, I follow their usage to indicate the highly pervasive nature of male domination. This is hard to pin down precisely because gendered hierarchical principles permeate most social institutions and many interactions and discourses. On balance, I feel that patriarchy is a term that will have work to do for some time yet in studies of smallholders.

Peasants as Patriarchs?

Can we speak of a 'peasant patriarchy'? As the previous discussion indicated, a variety of types of smallholding producers are encompassed by the term 'peasant'. Even analyses that attempt to wholly deconstruct the idea of a peasantry at the same time acknowledge the difficulty of doing so. Relatively self-sufficient petty commodity producing households, or middle peasants, may be neither proletarian nor incipient capitalists. In these households, family labour is crucial to household production as well as to its continuation as an economic unit. Even though today most rural people's livelihoods are diversified (see Introduction), where land reforms take place they increase the importance of household-centred production.

It is important to reiterate that enormous differences in gender regimes exist between and within societies. Variations exist in kinship and marriage systems, systems of marriage payment, religious beliefs, ideas about sexuality, family ideologies, and other factors. Certainly in a few systems, women have a high degree of economic autonomy, particularly in West Africa (Therborn 2004). Evidently, then, differences will exist in the status of peasant women in various social situations. A third wife of a middle peasant man in a Shona-speaking Communal Area of Zimbabwe, for example, will not be positioned in the same way as a north Vietnamese female household head. The situation of an older French woman, married to a prosperous peasant and living in a society which enjoins relatively companionate norms of marriage, will differ again. Even within one society, many differences among peasant women will exist. The situation of a poor peasant woman in Viet Nam is likely to differ if she is an ethnic majority *Kinh*, an ethnic minority person living in the Uplands, if she has trading opportunities, or according to the number of sons she has borne. This is to employ intersectional analysis and to indicate that an individual is always a classed, gendered, and ethnic subject. S/he will also be of a particular age and particular citizenship status, among others

Smallholding or peasant households do not necessarily produce one type of patriarchy, then, due to overall social and cultural variations. However, several features encourage relations of domination over women—or patriarchal gender regimes. One factor is the relative spatial isolation of smallholder communities and households. This means that customary or traditional gender regimes are more intact in rural areas. Another factor of relevance is the combination of production and domestic tasks in one household unit. In many societies, women's work in agriculture, craft production, and processing of crops is crucial, and peasant men normally direct the labour of wives and daughters. As noted in Chapter 1, smallholder farms are often seen as more efficient than larger agricultural units. The hidden ingredient here is women's work, especially that of wives. Although material exploitation can only partially explain women's subordination, a material basis for this exists within smallholder households.

Husbands and fathers also have an impetus to retain control over women's sexuality within peasant households. Male control of women's sexuality is of course not specific to smallholding cultivators. Two dimensions, however, are particularly marked. Control of women's reproductive capacity ensures reproduction of the economic unit itself through reproduction of children. Peasant households are not only social, commensual, and consumption units but are also units of production. This makes women's work and their sexuality of crucial importance, with resonance somewhat different to urban households based on wage labour. Further, control over women's sexuality has symbolic importance for male identity within many peasant communities. A married peasant man may control little outside his household, but he does usually have authority within the household and has a position of status as its head. Control of labour easily becomes control over bodies, especially those of women. Thus, any loss of control over women is particularly threatening and may result in loss of wider community status.

The comparative nature of this book provides an opportunity to consider continuities as well as different experiences in gender systems. The experiences of land redistribution, collectivisation, and decollectivisation have differential impacts on gender relations. Agricultural collectivisation threatens the position of the head of the smallholder household. Collectivisation has precipitated resistance, and in some cases this has been strong enough to undermine the direction of state policy (Stacey 1983; Wiergsma 1991). Conversely, land reforms on the individual household model have formed or else re-constituted peasantries. With redistribution of land, peasant households usually become much more viable economic units. A side effect of strengthening of the household as an economic unit has been that male household heads assume more power.

CONCLUSION

This chapter has concentrated on analysis of concepts relevant to study of gender and agrarian reforms. Most of these are contested and discussions have had to acknowledge their fuzziness. To an extent, this reflects great empirical complexities within rural economies and societies. The concept of patriarchy is hard to pin down in part due to its pervasive nature. Thus, some concepts employed are not definitive but must be used in a shorthand manner. They operate as ideal types as seen in the Weberian sense of models (Weber 2001). For instance, the ideal type of a peasant is a middle peasant or peasant household. In practice, however, a range of rural producers, agriculturalists, and semi-proletarians exist. Models do not approximate exact 'reality' but nevertheless provide a useful guide

I have stressed the importance of gender analysis of peasantries. This necessitates recognition that central concepts such as household or peasant must be deconstructed. Recognition of the unequal nature of power, work, and authority within peasant households and communities is also central to such analysis. Such inequality is often underpinned, in the last instance, by violence.

It is not possible, in my view, to speak of one 'peasant patriarchy' due to variations within and between societies. I do argue in a more qualified way, however, that smallholder economies tend to produce gender regimes in which men in a collective sense are highly dominant. This does not imply that women are passive, simply that their actions are often constrained. I argue in the next chapter that male domination has increased within many situations of land reform.

3 The Gendered Effects of Household Models of Land Reform

This chapter discusses how 'individual family' or household models of agrarian and land reform affect women's status and autonomy. It attempts a general overview of how processes of agrarian reform impact upon gender relations among agricultural smallholders. Land and agrarian reforms can take place either with redistribution to individual households or families or else can be collective in nature. The majority of reforms, especially land reforms limited in scope, redistribute land to households as units.

GENDER, PEASANTS, AND LAND REFORMS

Chapter 2 posited that the economies of peasant smallholders are often predicated upon the labour of wives. In such households, it is common for women to be subordinate. Much writing does not acknowledge this possibility and assumes that households are undivided entities in which the head of household manages on behalf of the family. Feminist critiques of the household have taken place separately from mainstream literature on agrarian reform. In six books on land reform in the 1990s, for instance (Christodoulou 1990; El-Ghomeny 1990; Prosterman et al. 1990; Barraclough 1991; Sobhan 1993; Thiesenhusen 1995), the topic of gender divisions either is omitted or else mentioned only in passing. Among the few general surveys of land reform published in the early twenty-first century C.E, Ghimire's (2001) does contain one chapter on gender relations. It has been left to a small group of feminists to analyse gender with regard to land reform policies. Carmen Diana Deere and Magdalena León de Leal and Bina Agarwal have perhaps the most extensive bodies of work on this subject.[1] A number of case studies are also available. The observations in this chapter are drawn from research dealing with gendered effects of the household models of agrarian reform. I discuss general effects of collective agrarian reforms for gender relations in Part II.

A note on sources and methods is in order. Here I concentrate on studies which document the effects of land reform upon women and gender relations and which are based upon *empirical* study. Works that critique policy alone are not included. Other chapters in the book explore different national contexts of agrarian and land reform where a more substantial literature is available. Because literature on gender and agrarian reform is relatively sparse in some instances, sources consist only of one or two articles. Thus, some of the cases discussed here are ones in which coverage is somewhat fragmentary. This chapter relies on twenty-six case studies of land and agrarian reform along individual household lines; they[2] are taken from Africa, Asia, and Latin America and include sixteen countries with brief reference to Viet Nam and to Zimbabwe, both discussed in more detail in later chapters. Ingrid Palmer (1985) has also written a hypothetical/composite case, drawing on studies worldwide.

Differences Among Women and Between Agrarian Reforms

Discussion of the effects of agrarian reforms 'for women' can lead to the impression that they are an undifferentiated grouping. Even where the literature does focus on women it often does not attend to divisions and differential social positionings such as ethnicity, 'race', social class, generation, sexual orientation, and marriage status. The extent to which gender predominates over other positions and identities depends to some extent on particular instances (Anthias and Yuval Davis 1983). Theorists of intersectionality (Brah and Phoenix 2004; Yuval Davis 2006; Anthias 2006), as noted, point out that gender is always inscribed in class, ethnic, and national positionings. Likewise, other categories such as ethnicity and class are inscribed within gender positionings.

Just as there are differences between women, so also agrarian and land reforms do not fit into a single category. The extent and type of agrarian or land reform remains of much importance (see Chapter 1). To give one example, the Philippines case is one of very limited, 'conservative' land reform in which peasants gained little power (Borras 2005). Support for land reform was inadequate, as was evident in several ways. The designated land covered only certain crops (ie rice, corn), and more importantly, landlords were able to ignore government-issued certificates of transfer and thus to demand continued payment of rent (Lindio-McGovern 1997: 49–50). Only husbands' names appeared on certificates. Neither men *nor* women, however, were able to gain control over the land they cultivated. This contrasts with cases in which large amounts of land were redistributed and which had more egalitarian aims. Hence, the general context of land and agrarian reform remains crucial. Additionally, each specific case of land reform contains aspects which are beneficial to [most] women and other aspects which are detrimental, in that they tend to undermine women's power and autonomy.

Individual cases are important, but so are comparisons which are able to highlight continuities. I will summarise and discuss the main factors that emerge when considering individual tenure land reforms in gendered terms. Such models have had markedly similar results. This is striking, given wide variation in geographical regions, in extent of reform, in the status of women, in culture and religion, and in kinship forms.

Agrarian reforms and the household head

Household models of reform usually exclude married women from the redistribution of land or subsume them under the husband's title. Only people deemed to be 'household heads', nearly always men, are granted land. Palmer wrote, 'A woman's access [to land] is akin to that of a bonded labourer' (1985: 30). In effect, male household heads may be confirmed as a new class of small landowning farmers, particularly where previous tenure was communal. In Chile, for instance, a limited land reform took place under Eduardo Frei (1964–1970) and then a wider agrarian reform under Salvador Allende (1970–1973). These benefited 20 per cent of the rural labour force (Deere 1986b: 190). Under Frei's Christian Democratic government, the reform strongly favoured the *inquilinos* (permanently resident agricultural workforce). Criteria specified that beneficiaries be permanent full-time workers and married heads of household. Nearly all women apart from widows were excluded, as well as many male peasants. In Iran, male household heads became beneficiaries in the 1960s land reform. A new sphere of gender asymmetry, land ownership, opened up (Razavi 1994: 613). As had been assumed by planners, male cultivators were able to exert control over their wives' labour.

This aspect of agrarian reform policies is crucial and sets the basic parameters within which such policies must be analysed. Allocation of land to household heads rather than all adults in the household means that wives start out structurally disadvantaged vis-à-vis husbands. Agrarian reforms may bring other, possibly advantageous, changes. However, they do so within a framework that potentially exacerbates female subordination.

Widows with dependent children are allowed to hold land within many programmes, and this is potentially of much benefit. For instance in Honduras, Iran, Chile, and Tanzania, female household heads—defined as women without adult males present—were included. In practice, however, few benefited. In Honduras, 27 per cent of households were female headed in the 1980s. However, a survey of 32 *asentamientos* (land reform settlements) found that there were either no or else very few female household heads who were beneficiaries (Safilios-Rothschild 1988: 217). In a few cases such as in West Bengal, land redistribution meant a *loss* of widows' land to male beneficiaries (Agarwal 1994b: 281).

GENERAL CONSEQUENCES OF LAND REFORMS: LAND RIGHTS, WORK, SERVICES, AND INCOMES

Household models of agrarian or land reforms often mean a loss of women's land rights. In some areas or regions, particularly in much of sub-Saharan

Africa, married women customarily have access to a plot of land on which to cultivate 'women's crops'. Usually these are food crops such as groundnuts or vegetables. This land is normally part of the husband's communal allocation, and he usually administers the 'wife's' land. In the past, it is likely that nearly all wives had such an allocation. A husband could not leave his wife without means for subsistence; in any case it was (and is) the wife's responsibility to provide for herself, her husband, and her own children. Hence women usually cultivated crops for consumption, although sometimes they were able also to cultivate crops to trade or to sell. Thus, women often had rights of disposal over crops produced on their land. This right was eroded with changes under colonialism and then capitalist development but still usually exists.

A number of sub-Saharan African studies report that women lost rights to their customary plots with land reform and land resettlement. In the Mwea scheme, Kenya, a woman could no longer be sure of obtaining her own food crop garden and so depended on the husband's goodwill to meet her obligations which remained unchanged in the new setting (Hangar and Moris 1973). The shortage of land on which to grow crops was seen as the biggest difficulty of life in the new scheme (Hangar and Moris 1973: 229). In the Tanzanian *ujamaa* village studied by McCall, women were allowed to be village members in their own rights but were not allocated their own land if married. Legal and cultural changes left women dependent upon men in the household as well as on male village leaders (McCall 1987). Thus, for married women the lack of legal title and of physical control over land was the most fundamental cause of their subordination. Their previous customary rights had been eroded and they were left worse off (McCall 1987). Ethiopian women lost land rights for somewhat different reasons. Amharic women *could* inherit land bilaterally. The agrarian reform abolished inheritance of property and with it, the main means of acquiring land.[3] Perhaps unintentionally, this left women more dependent upon husbands (Tadesse 1982).

Agrarian reforms often increase work burdens for all. A number of accounts report a disproportionate increase in women's work burdens. Initial preparatory work is arduous, and there is more land to cultivate, more pressure on labour generally, and an urge to make a profit. Thus, in the Mahaweli scheme, Sri Lanka, women's responsibilities and work on the farm increased (Lund 1978). Following the transformations from a collective to household economy in a number of societies, women's work also increased due to the withdrawal of state services (Liljeström et al. 1998; see Part II). In yet another case, in Tanzania women's work burdens increased with villagisation under ujamaa. This took place for several reasons: farmers' fields were further away; women had to walk further for water and fuel; and preparation of initial homeplots, unusually, fell to women. Furthermore, without inputs such as fertilizer, only additional human labour could maintain production. Women found themselves pressured to work on husbands' fields (McCall 1987). In southeast Iran, a very similar process took place (Razavi 1994). Possession of more land and orientation to cash crop production intensified wives' labour and increased their workloads. This process was even more marked in the previously mentioned Kenyan scheme in Mwea. There, the gender division of labour broke down with resettlement. This loosening of the gender division of labour, however, was to women's detriment. Men still avoided 'women's' tasks such as taking bananas to market, weeding, carrying water, and cooking, but women had to take on many male agricultural tasks (Hangar and Moris 1973: 225). Likewise, in an Upper Volta (now Burkina Faso) resettlement scheme, women's work increased. This was due to reorganisation of production and reproduction (Conti 1979: 75). Women lost control of work that had traditionally been theirs. They also became more dependent upon husbands (Conti 1979: 88). In general, then, the extra burdens of work with agrarian or land reform often fall particularly upon women.

Issues of social class within land reforms also impact upon women. Land reforms usually increase class differentiation among peasants while also decreasing that between them and large landowners. In Chile, the initial land reform beneficiaries in the 1960s, as noted, were permanent agricultural workers. This excluded male temporary workers and *afuerinos* (those not residing on the estates; Garrett 1982). The initial beneficiaries gained a good deal of status in class as well as gender terms (as will be seen) from the distribution of land and cooperative membership.

The class dimensions of land reform often exacerbate divisions *between* as well as within households (Razavi 1994). Studies in Ethiopia (Tadesse 1982) and Mexico (Young 1981) found that wealthier peasant women's labour is least essential to the household, since labour is more likely to be hired in. In some areas and cultures, a notion of dishonour attached to women's public field labour may also exist (as will be seen). Another pattern noted in Mwea, Kenya, (Hangar and Moris 1973) and Zimbabwe (Jacobs 1989, 1995) is that where polygyny is common, wealthier peasant men may marry a number of wives to meet labour needs. Middle peasant women's labour tends to be most important for smallholding households who hire in few outside labourers (Jacobs 1989, 2000b; Chapter 2). I found in Zimbabwe that middle peasant women were more subordinate in the household than poorer and wealthier strata. This may be linked to the need for their labour. Similarly, middle peasant women in India have been found to have less influence compared with wealthier and poorer producers. Here, dowry and enhanced status associated with not working outside the home are significant (Robert Chambers, personal communication, 1988; see also White 1992). Agarwal (1994a) argues that middle and wealthier peasant women are less likely to oppose husbands than are poor peasant women because wealthier peasant women have more to gain through identification with the household.

Examination of poorer peasant households commonly finds that women are most likely to labour for wages outside their households. Even if some stigma is attached, bringing in wages may increase women's power within households (Dwyer and Bruce 1988). Peasant women in poor Honduran households were most likely to control household income in a region where female control was seen as customary (Safilios-Rothschild 1988). Middle and poorer strata women have the heaviest work burdens, especially when households are female headed or hire little outside labour. In Ethiopia, women's work increased with land reform and market domination, *especially* for poor peasant women. As well as new tasks, they had to work hewing wood, drawing water, and fulfilling other customary duties (Tadesse 1982). Many, although not all, female-headed households are poor, as is now commonly observed. Their poverty also reflects gender disadvantages that are also intertwined with social class. Women-headed households are often unable to command labour and also lack access to other 'inputs' such as credit and new technologies.

Many agrarian and land reforms are marked by poor provision of services, especially in early stages. It is common to find that schools and clinics are in the process of being constructed. Similarly, though access to potable water is often a government aim, women frequently have long distances to travel on foot, especially to fetch water and fuel. Poor communications also exacerbate problems of marketing. Conditions on new land resettlement and land reform areas are often arduous, especially because of poor service provision. In Mwea, Kenya, for example, many women disliked living in the irrigation scheme partly because of the difficulties of obtaining fuel, the pollution of water, and other health difficulties. However, a women's self-help scheme existed, and it opened several shops and kiosks within a few years (Hangar and Moris 1973: 220). In Upper Volta, the outcome was more negative. Women on resettlement schemes did find that some schools and health posts were available but that wells were few and far away. Grain mills were eventually established, but these were operated mechanically as well as for profit. Because milling of grain became a productive activity, it was taken over by men. Men were also advantaged, unlike women, by having access to credit (Conti 1979: 86). In Honduras, few training opportunities were available for peasant women within the agrarian reform (Safilios-Rothschild 1988). In contrast, services in Tanzania such as access to shops, roads, and health services *improved* with villagisation (McCall 1987: 197). Overall then, a mixed picture exists concerning provision of services, infrastructure, training, and other backup for land and agrarian reforms on which the success of projects often depend. Gaps in services may be particularly difficult for women as they are usually responsible for providing food, fuel, water, and healthcare for their families.

Do land and agrarian reforms raise general living standards? A main aim of policy is usually to raise peasant subsistence and income levels through enacting reforms. Even studies that hold that they exacerbate class and gender divisions find living standards and food security are often improved. Indeed, if this does not take place, then the agrarian reform will usually be termed a failure. Tadesse (1982) reported that for Ethiopia, resettled peasant women were enormously relieved that they did not have to worry about absolute food security—where the next meal was coming from. In Libya, the Kufira settlement raised living standards and particularly, standards of housing (Allaghi 1984: 138), although other gendered effects of the reform were more negative. In Andra Pradesh (India) stability and food security were seen by women as an important marker of success of redistribution of land (Raghunath 1996: 361). Increased workloads, as noted, are often necessary to secure increased production.

Do better living standards and higher incomes filter down and increase women's own incomes? According to most studies, married women typically lose control of their 'own' incomes while that of the household head rises. In this respect, this factor parallels and is related to that of female loss of control over land. Nearly all studies report similar results. However, cases that are aberrant in this respect such as Zimbabwe also exist (see Chapter 8).

Women lose income for a number of reasons including loss of access to raw materials and to land; loss of economic niches through relocation (ie trading and marketing opportunities); the need to travel very long distances; and loss of personal contacts. At the same time, men often acquire monopolies of new cash-cropping opportunities unavailable to women or else available only through male mediation. Women in Honduran asentamientos usually earned less than they had previously within villages. Where they did receive any land, this was usually of poor quality. (See also Momsen 1988 on Caribbean peasant agriculture.) Opportunities for women were either static or had been reduced. Overall, agrarian reform in Honduras depressed women's incomes and *increased* inequalities between spouses (Safilios-Rothschild 1988). In Mwea, women were paid in kind for their work, but men were paid in cash, increasing inequalities between the sexes (Hangar and Moris 1973: 241). Men often acquired monopolies of new cash-cropping opportunities in land reform. When men gain in wealth and economic opportunity, their power vis-à-vis women in the household often increases. That land reform processes may lower wives' incomes at first seems counter-factual. It seems, however, that in many cases raising household income has meant raising that of men. This has not always accompanied wives' control over their *own* incomes.

FAMILY, KINSHIP, REPRODUCTION, AND DECISION-MAKING

Agrarian and land reforms affect parameters of household relations other than those concerning living standards and production. They often have profound implications for kinship, family, and internal household relations.

Various changes in family forms and lineage structures are often concomitant with agrarian reform. Growth of the nuclear family is important and relates to the fact that reforms often involve relocation as well as to processes of 'housewification'. People who are moved to unfamiliar areas with unfamiliar people may retreat emotionally; they may turn inwards towards the 'couple'. Colson noted (1960) that husbands and wives may come to rely more upon one another with resettlement. This implies dependence of the husband upon the wife as well as the wife upon the husband. Libyan planners, themselves mainly urban and middle-class, developed a model of housewifery to which rural women were unaccustomed. Rather than providing agricultural skills within land reform schemes, they shunted women into cookery and handicraft programmes (Allaghi 1984: 139). This was ironic as previously women's economic roles had been extensive: collecting water; basketweaving, aspects of farming; and herding sheep. Within land reform areas, men did most of the farm work, and women's roles were reduced to housework. Although housing and living standards improved, women became much more dependent upon their husbands. The Volta Valley Authority scheme (Upper Volta) entailed setting up relatively small holdings with nuclear families settled on them to ensure a better 'redistribution' of reproductive labour power (Conti 1979: 85). Sri Lanka presents perhaps the clearest picture of growth of nuclear families. In the Mahaweli study, husband and wife became a more self-contained unit with their children and more isolated from kin (Lund 1978; de Silva 1982).

A number of changes in marriage relations occurred among Amharic people in Ethiopia with agrarian reform (Pankhurst 1992). Some were due to impoverishment as well as to agrarian change. There was, for instance, a reduction in the ceremonial aspects of marriage (of which various types exist) and greater informality and say over choice of marriage partner. Pankhurst also found later marriages and choice of spouses within a closer geographical range than had occurred previously (1992: 125). Issues of land allocation with the agrarian reform tied women into marriages in a manner that had not been customary within this part of Ethiopia. Unusually, Amharic women previously had a degree of leeway in leaving unsatisfactory or abusive marriages, although men mainly owned the house and retained it. Reform also led to women losing rights to inheritance which, unusually, they possessed. Ethiopian agrarian reform before 1991 followed the common practice of allocating land to men (Tadesse 2003: 85). Additionally, women were usually excluded from the peasant associations, made up of male, usually wealthier, peasants (Pankhurst 1992: 25). As a result, following land reform, marital status rather than descent as well as attitudes of the peasant association, determined whether a woman could claim land. Thus, women lost rights as well as the degree of flexibility they had had in leaving marriages because land rights became attached to marital status.

Still another scenario has occurred in 'left' developmental contexts in Kerala and in West Bengal. Agrarian reform, in raising the incomes and status of peasant households, has also fostered aspirations for improvement in caste and class status, especially among those who were previously landless (Kodoth 2005). In West Bengal, these have been in conflict with women's work outside the home, particularly in agricultural fieldwork. Such work is associated with low caste and class status and is seen as shaming. It is accompanied by taboos on women's use of ploughs. Thus, women such as widows who have land often have to hire in labour (Agarwal 1994a). Increased prosperity has been associated in this case not with emancipation of women but with restrictions on widow remarriage, female seclusion, and increases in dowry payments (Kodoth 2005: 2545). Large dowry payments in turn have been associated with alienation of land. This may include the very plots gained through land redistribution being lost in order to raise revenues for daughter's marriages. Thus, daughters may be seen as an economic burden.

In West Bengal, attention has been paid to women's land rights at the policy level. A 1992 state directive decreed issue of joint titles to spouses where possible. Various studies found that in practice few women had received land. A statewide survey of land reform beneficiaries indicated that less than 10 per cent of plots were in the joint names of spouses, and where they were, wives were often unaware that their names had been recorded. Only 5 per cent of titles were in women's names alone (Chakraborti 2003, cited in Kodoth 2005). Another study of 800 households found that *no* wives' names had been included. Peasant unions in this case had refused or failed to assign land to married women, making exceptions only in cases of perceived distress (ie for women without adult sons who were divorced or deserted); (Gupta 1997, cited in Kodoth 2005). The few women who had received land in some cases alienated this for daughters' dowries. Kodoth explains,

> The highly skewed distribution of property rights in land combined with the growth of dowry, leaves little room for dispute that the agrarian reforms have reinforced a patrilineal framework (2005: 2547).

When the nuclear model reinforced by land reform is also of a monogamous family, one or several wives may be completely left out or deserted. A man may be able to register only one wife, as occurred in Nigeria (Fapohunda 1987). In Ethiopia, registration of one wife meant that others were left behind, in a context in which 10 per cent of marriages were polygamous (Tadesse 1982: 214). The wife (or the only legitimate wife), without access to land or to her own sources of income, became more dependent upon the husband. Usually this occurred at a time when she was separated from wider kin networks. A silence exists concerning the other wives, but it seems fair to assume that few would have prospered. General disruption also appears to render women more vulnerable to divorce, which in turn may lead to impoverishment.

The studies cited note different opinions concerning the impact of nuclear family structures. In reinforcing a nuclear family structure, agrarian reform may diminish the power of extended families. For this reason, as well as because of spatial separation, links with wider kin may be weakened or lost. How these processes are viewed depends considerably upon the prior kinship and marriage system. Many women prefer nuclear families to patrilocal patrilineages. In Viet Nam, women preferred the smaller family system under land reform because it attacked the power of husbands' patrilineal relatives, especially that of the mother-in-law over younger wives (Gammeltoft 1999; see Chapter 6).[4] The nuclear family in this instance enhances wives' limited power. In Zimbabwean schemes most women also preferred to be free of the power of husbands' patrilineage relatives (see Chapter 8). In Sri Lanka, the nuclear family meant that women could become more assertive in marriage, away from in-laws (Lund 1978). Also positively, De Silva's Mahaweli study found that husband and wife had become more like economic partners than had been the case before land reform (1982).

The nuclear family model is often double edged for women, meaning more interdependence, intimacy, and potential influence but also a growth in surveillance and control by the husband. Thus, in Mexico after reform, wives' scope for decision-making was reduced, due to loss of women's land, greater surveillance by the husband, the increase in male authority and *machismo*, and women's greater confinement to the role of 'housewife' (see Brunt 1992; Chapter 7). Settler women in the Mahaweli scheme, Sri Lanka were also disadvantaged, becoming more dependent upon husbands because they were further away from their own kin. They came to depend upon the husband even for basic information and did not always receive this. Women's economic power had been constrained previously and was then further reduced. In particular, they took little part in decision-making concerning agriculture (de Silva 1982: 143). Wives in Mwea became more dependent on husbands for provision of basic foodstuffs and other roles became more circumscribed (Hangar and Moris 1973). In Libya, in theory, rural women could participate in training schemes set up during resettlement and land reform; however, in practice they became more isolated within housewife roles and took less part in decision-making either outside or inside the home (Allaghi 1984). In the initial land reform in Chile, gender hierarchies were strengthened. Women's economic dependence upon men was reinforced as was men's sense of authority over wives (Tinsman 2002: 172).

An underlying reason for the diminution of women's decision-making power in agrarian reform lies in strengthening of peasant family units and men's power within these. Previously, impoverished peasant or landless men were likely to have found exercise of authority more difficult for two reasons. One is that landlords exercised authority over peasant men and sometimes over women peasants. A possibly more important factor was that most poor peasant women brought income into the home, and the husband depended in part upon this.

Women's increased 'encapsulation' within the home has sexualised dimensions. In Chile, women's dependence increased as they had less need to work outside the home. However, land reform also bolstered the pride of male beneficiaries. Men's authority in the home increased as they were eager to display a reinvigorated masculinity, previously denied them as *peones* and subjects of the landlord. Many men incessantly policed the parameters of female domesticity (Tinsman 2002). Later, under Allende, attempts were made to democratise the agrarian reform. Wider groups of men as well as more women were allowed to join the new agrarian reform centres (CERAs). In practice, few wives participated because of their long hours of work. However, an equally important factor was male opposition: one of the most frequent complaints against the CERAs was that they *permitted* women to be members (Garrett 1982).

Another aspect of land reform's influence on family structure concerns the size of households. Family size has economic and religious as well as sexual dimensions. Decisions over family size may be affected by any decrease in women's household influence. Land reforms may increase the pressure to reproduce children, given the importance of family labour in agriculture (Palmer 1985). This was an effect, for instance, of the 1990s redistribution of land in Viet Nam, *despite* the context of restriction on family size (Gammeltoft 1999). Pressures arising from a peasant-based household economy, then, may conflict with state goals to reduce family sizes. Wives' room for manoeuvre in this crucial area has been reduced in at least some land reform schemes.

Beyond the family realm, there are also doubts about whether reforms promote cooperation. Many reforms entail resettlement in new villages and place people in unfamiliar spatial arrangements. Although this could provide a basis for cooperation, reports suggest that mere proximity does not automatically have this effect. Rather, people often feel crowded, displaced, and suspicious of others. Where belief systems stress the power of ancestors, such suspicion may be greatly heightened by worries about living on other lineages' ancestral lands. In Mwea, Kenya, new villages were very compact and followed new settlement patterns (Hangar and Moris 1973: 220). This, along with shortage of land for cultivation of food crops, caused stress. In Sri Lanka, settlers were also 'beset with stress' in the first years (de Silva 1982: 140). Thaya Scudder (cited in de Silva 1982) put forward a theory of multidimensional stress: physical, psychological and socio-cultural. Scudder held that such stress nearly always occurred with removals and resettlement and most commonly affected women. Mwea women, for instance, said that they felt very lonely in the new resettlement. Across the world in Chile, too, many wives felt that land reform had resulted in greater feelings of isolation. Their work in and near the home increased and there was no longer need for community activism in land struggles (Tinsman 2002).

Cooperation among neighbours might grow, however, with familiarity. McCall outlines a more optimistic scenario, in terms of growth of communality—although this case involves more communal elements than others discussed here (see Chapter 8). Villagisation in Tanzania presented many drawbacks such as increased pressure for peasant producers, especially women. (1987: 212). Villagisation, however, also helped to end the social isolation of women, making them more aware of shared problems and of gender issues through associating outside their own families (McCall 1987: 213).

CONCLUSION

Some of the effects of land reform outlined in this chapter are 'positive' for the lives of peasant smallholder women. Prime among these is increased household income. Assuming some degree of redistribution within households, this benefits women within families, albeit usually less than men. Increases in income are likely to have some impact on food security, the principal goal of land reform. This is not automatic, however, as increases in income can also be used in a variety of ways, and income is usually under male control (Blumberg 1995). Given the importance of food security to people's well-being and to their very lives, income is of overriding significance for many women.

Successful agrarian reforms provide good infrastructural support. Where roads, shops, and clinics are constructed, boreholes sunk and agricultural inputs provided, these too usually improve women's lives. Again, however, such benefits are unlikely to be gender neutral. Provision of services to land reform areas may be slow or sometimes nonexistent, making the first years of land reform and resettlement arduous.

Changes in family forms to nuclear structures are also perceived as beneficial by many women. This is especially so where larger lineage-based patrilocal households had existed prior to land reform. Nuclearisation of families usually increases informal influence a wife might exert. Crucially, the power of in-laws over her lessens. Where polygyny exists, nuclear family models mean that the first (or senior) wife ordinarily gains, while secondary wives are excluded. The growth in nuclear family forms often occurs at the same time that people move into new communities. New extra-kin links provide possibilities for women. However, these may take time to develop or may be impeded by communal distrust and hostility.

These positive aspects of agrarian reform are important. Other effects may be more detrimental for women or else for some groups of women. Firstly, women often lose rights to their own land for food crops, where these have existed. Relatedly, most studies report a decrease in income that women control themselves. This occurs *even if* household income is raised. Land reforms commonly increase women's work burdens, although men too may have extra work. Additionally, married women may be under more pressure to bear children. The nuclear family model may afford wives a degree of *informal* power; however, where they are pushed (or pushed back) into the household in housewife roles they often lose autonomy.

Within agrarian reform policy and practice, some provision for wid-ows and deserted women is usually made, though as stressed here, land is nearly always allocated to a male household head and so married women are disadvantaged. Even in rare cases (e.g. West Bengal) where land titles or permits should be issued in names of both spouses, the law is circumvented in practice. Allocation of land in men's names increases their household and community power and renders women's rights less secure. In nearly all cases, a divorced woman is expected to leave the 'husband's' land and house, so participation in land reform may be contingent upon marriage. In case of widowhood or divorce, most women's situations deteriorate mark-edly, particularly without access to land.

It is the aim of land reform to constitute or reconstitute peasant house-holds and to ameliorate peasants' conditions of life. Sometimes, wives may be able to exert influence within such households, but they are rarely contexts in which women have significant autonomy or hold overt power in their own rights.[5] External agencies such as states, local governments, the World Bank (Manji, 2003a, 2003b) and political parties may fail to disaggregate the peasant household. Such policies result in perpetuation of household gender subordination. Likewise, populist and other accounts wishing to support peasant movements are often reluctant to disturb the unity of households or to point out women's disadvantaged place within these. The argument is that, after all, male peasants too suffer oppres-sion and impoverishment. In this context, the strengthening of the middle peasantry is significant within land reform (see Chapter 2). These are the households most dependent upon female labour performed *inside* the economic unit.

Repeasantisation within land reforms has meant that women have often lost autonomy. By 'autonomy', I mean not complete independence; few human beings have this. Rather, I refer to space to negotiate; or, to use Kandiyoti's term (1988, 1998), to 'bargain'. It is worth noting that often women's bargains are in pursuit not only of their own interests but also include those of children (see Elson 1995). Many wives have benefited materially through agrarian reforms. This is important to acknowledge, but there has been a price, the loss of room for manoeuvre inside and out-side their households. Land and agrarian reforms have often benefited hus-bands at their wives' expense.

Part II
Collectives and Decollectivisations

4 Gender and Agricultural Collectives

Soviet-type Economies

Twentieth-century political revolutions, particularly those influenced by the socialist ideologies of the Russian revolution, have envisioned land reforms as a key policy strategy for addressing economic development and social change. Within these, collectivisation of agriculture has been a recurring policy, though not always an enduring one. Studies of collectivisation, while recognising their class implications, have generally overlooked implications for gender relations. In this chapter, I therefore discuss case studies of those states where researchers have made some attempt to incorporate gender. This chapter begins with the Soviet Union and then discusses the ex-Soviet state of Uzbekistan. It then follows with examples from two formerly Soviet-type European states, Bulgaria and Hungary, which were pressured to collectivise (Meurs 1999a). Lastly, I examine the transformations within Cuba, influenced not only by its political relations but also because it represents a different agricultural economy and historical relationships.

The 'foundational' status of the USSR and its experience meant that many other states and movements within the Soviet sphere of influence established similar organisation of agriculture. The Soviet approach to economic development stressed industrial production and saw agriculture as simply contributing to it through providing grain supplies and revenues. Collectivisation was not simply an economic strategy but also a political one, aimed at proletarianising the peasantry, which was seen as a potential source of opposition to socialism. Soviet-style governments also attempted to regulate gender relations within and outside families. Here, policies drew on Engels' ideas in *The Origins of the Family, Private Property and the State* (1972). He saw women's oppression as caused by bourgeois (or feudal) family and kinship structures which gave men economic power over women. Thus, the basis for emancipation lay in bringing women into production on a similar basis to men. Unfortunately this interpretation ignored women's existing contributions to production and the economy and assumed that no other significant bases for women's subordination existed. Nevertheless, this model did pay some attention to equalising women's position within the economic sphere. Importantly, family law was usually reformed, including outlawing of practices such as polygamy and forced marriage.

The importance of these steps should not be minimised as they had crucial impacts on women's and men's lives within households. They could not, however, encompass aspects of women's subordination unexplained within this frame. Experiences of collectivisation which deviated from an Engels-type model are very few.[1]

APPROACHES TO COLLECTIVISATION

Across contexts, a variety of forms of agricultural production cooperatives and collectives exist. The most collective of such units are state farms, in which the state owns all land and inputs and workers are fully proletarianised, receiving wages as in industry. Often, however, households are permitted to hold garden plots, as in the USSR and Bulgaria (Meurs 1997). In contrast, cooperatives are forms in which land is held by peasants who operate in a more independent manner, although needing to meet certain targets. Other forms may employ collective labour in only part of the production process, retaining independent peasant units with a degree of cooperativisation of production.[2]

For a number of reasons, most socialist countries went beyond voluntary cooperativisation and on to full collectivisation (Nolan 1988; see also Jacobs 1997). Firstly, an independent peasantry was considered to be a political threat on the assumption that peasant consciousness was individualistic and likely to take decisions serving personal short-term interests. This was contrasted with the proletariat, viewed as communally minded and potentially revolutionary (Nolan 1988). Full collectivisation proletarianises the peasantry, turning it into an agricultural wage labour force. It aims to curtail class polarisation within the peasantry, though some have argued that cadres in effect formed a dominant stratum within collectives and that a new class emerged (Nolan 1988: 35). Collectivisation was also seen to have social advantages, especially for poorer peasants, by providing full employment for most men and often for women as well. It also provided schools and clinics, benefiting all the collective, including women.

Governments also saw advantages in the collective model. From one perspective, it provided a way to market farm production and to siphon off rural savings to finance non-farm investment. By so doing, however, leaders often failed to reinvest in collective farming, thereby damaging the motivation of collective members. Collectives were also seen as attaining economies of scale in production and as offering effective routes to diffusion of new techniques efficiently, due to their centralised nature (see Chapter 1). What these approaches overlooked were the potential inefficiencies of large-scale operations. Because much agricultural work is sequential and farm workers or peasants shift tasks frequently, permanent job specialisation is not possible or always desirable. An intricate interweaving of tasks characterises most farming. Within this process, women's labour and ability to combine different tasks is often crucial. Further, given its dispersed nature, farm labour is very hard to supervise. To address these conditions, many collective strategies eventually contracted out land to work groups or even to households. The model also failed to realise that while some crops may be suited to large-scale farming, others are more effectively managed on a small scale (Wittfogel 1963).

Although these Soviet-inspired approaches have been pervasive, and later recognised as failing to lead to anticipated results, comparative analyses suggest variables that play an important role in the economic viability of collective production (Meurs 1999a: 24). Collectives performed best when they were small- to medium-sized units farmed by a stable group that exercised a degree of social control over itself, avoiding the distrust that commonly existed between collective officials and members (Nolan 1988), and where mechanisms existed that could link income to individual work effort. When these conditions were met, a significant number of peasants would voluntarily join cooperative production units. Under some conditions, even larger scale collective agriculture could produce rapid growth (Muldavin 1998). For example, in Bulgaria between 1948 and 1958, growth proceeded at 3 per cent per annum and at 5 per cent between 1958 and 1963, although thereafter growth rates declined (Meurs 1999a: 18).

Despite the assumption that women's participation in production would provide liberation and equity, collectivisation was carried out with little explicit attention to any implications for gender relations. This was an odd omission, since collectivisation directly and immediately affected the form and roles of households and the lives of most peasant women were very much framed within households (see Chapter 2). In other words, these policies were carried out as if gender neutral. Yet, collectivisation did have profound implications for household and wider gender relations. In a sense, this was recognized by the common rumours and counter-propaganda to the effect that agricultural collectivisation would also mean collectivisation of women. Nevertheless, the anticipated dramatic change in gender relations mainly failed to materialise. To the extent that collective forms took economic functions away from smallholder households, they affected their workings and the roles of men and women within them. In particular, collectivisation took some (or all) direction of agricultural labour from the head of household to another body such as the collective committee, to cadres, or to another administrative authority. The separation of reproductive and productive labour within the household and outside it in the collective, affected relations and the workings of power within the household.

THE USSR AND POST-SOVIET RUSSIA

Soviet Collectivisation

This section discusses collectivisation in the Soviet Union as well as some aspects of decollectivisation in post-Soviet Russia. The Soviet case was the first socialist collectivisation, historically, and it (and its inefficiencies) often served as either a model or else a warning for others.

After the fall of the czar in the 1917 revolution, Russian peasants appropriated land for their own use. Initially, the Bolsheviks supported the redistribution of land but subsequently adopted a policy of nationalisation. Then, during the Civil War of 1918 to 1920 following the revolution, the Bolshevik government enforced a policy of grain requisitions and rationing. This precipitated a response by the peasantry of reducing sown areas: hunger and food rationing resulted. As a result, and with the need to secure support from the majority of the population in mind, the New Economic Policy (NEP) enacted in the 1920s allowed peasants to lease land from the state. Forced requisitions were curtailed and producer prices raised. From the late 1920s, however, those in favour of rapid industrialisation won the political argument.[3] With the growing absolutism of Stalin's power, opportunities for flexible responses disappeared. Gradual collectivisation and cooperativisation were replaced with a rapid and forced process while the strategy of industrialisation concentrating on heavy industry, financed by revenues raised from the peasantry and rural residents, was adopted.

The collectivisation process was conducted with great brutality under the guise of *dekulakisation* (a campaign against wealthier peasants who were targeted, expropriated, exiled, and frequently attacked or killed). Yet, as Lewin (1968) has argued, state actions were also directed against the *mass* of the peasantry, the (relatively) self-sufficient middle peasantry, not only the much narrower stratum of wealthier peasants. Massive peasant resistance took place (Viola 1996), including riots, overt and public assaults on collective officials and on poor peasants; slaughter of horses and draught animals; and breaking of machines, including tractors. While commonly resistance took the form of covert assaults, murders, and other acts of retribution on collective officials (Viola 1996), another piquant form of resistance was mass letter writing campaigns, including many letters from women.

The situation was no less than one of a civil war in the countryside, at a time in which peasants constituted approximately 80 per cent or more of the population. Women, too, participated in *bab'i bunty* (collective protests), which were viewed as hysterical manifestations of irrational women who themselves lacked agency. For the women, a major issue concerned communal ownership of livestock, especially cows. Cows were traditionally under female supervision and had spiritual connotations (Bonnell 1997). Peasant women were also mobilised by the attacks on religion that accompanied collectivisation and by rumours of communal wife sharing. Perhaps because of the prevailing view of the *baba*, these protests were treated with relative leniency. That is, the women were not charged with counter-revolutionary crimes, although '*kulak*' women would be deported along with their families (Bonnell 1997: 109).

While the majority of peasants opposed collectivisation, a minority did support the process, and these included a number of peasant women. This was perhaps because the government promoted women's participation and leadership in the collectivisation drive. For instance, it did its best to open up the ranks of relatively privileged tractor drivers to women (Fitzpatrick 1994: 10). Women were sometimes promoted to positions of authority in the *kolkhozy* (1994: 181). Similarly, the (later) Stahanovite[4] movement to boost production encouraged a rhetoric of women's liberation and had some success in encouraging young rural women to join in (Buckley 2006).

Evidence of the complexities of the early period of collectivisation and of resistance to it can be found in the political posters of the period, particularly important means of communication because many people remained illiterate (Bonnell 1997). Some displayed strong *kolkhovnitsa* (strong female collective workers). (See Figures 4.1 and 4.2.) Conversely, the negative images of kulaks were usually of men (Fitzpatrick 1994). Such posters, as well as being propaganda, perhaps also acknowledged that women were a group with something to gain and recognised that women workers, even in the 1930s, made up the majority of the rural workforce (Fitzpatrick 1994: 182).[5]

For some women, especially young peasants and older women (often, those who had been deserted) who headed households, self-assertion as workers was sometimes accompanied by flouting of male authority at home. Many male and some female peasants, however, found this offensive (Fitzpatrick 1994: 12). Indeed, peasant women who sought to take advantage of opportunities by becoming tractor drivers or kolkhoz officials were far more harshly judged than men behaving in similar fashion. Female Stahanovites were often treated as collaborators deserving public humiliation or worse (Fitzpatrick 1994: 12) and were generally viewed much more harshly than men (Buckley 2006). In becoming peasant activists, such women infringed gender norms as well as community norms of solidarity against Soviet 'outsiders' and they were targeted. Thus,

> Women activists were particular targets of peasant wrath, receiving treatment ranging from curses to arson and murder. . . . In the mid-1930s, collective farm women Stakhanovites faced persecution, beatings and rape by fellow collective farmers and by relatives (Viola 1996: 228).

Despite hopes of advancement, few women in fact held office on collective farms at this time. In 1937, for example, a group of women in Ostrov *ra'ion*, *Pskov oblast* (administrative district) wrote to the local newspaper to protest that there were no women chairs of Soviets and only two kolkhoz chairs in the region (Fitzpatrick 1994: 182). Despite such recognition, the common view among peasant men and many women was simply one of disapproval. Women in such roles, and they were rarely chosen, often faced marginalisation.

Figure 4.1 'Peasant woman to the kolkhoz!' (*Krest'ianka idi v kolkhoz!*), Moscow, 1932. Publisher Gosudarstvennoe Izdatel'stvo, Moscow. Acknowledgments to Hoover Institution Archives, Stanford University, California.

Overall, peasant resistance and conflict related to imposition of collectivisation resulted in disaster for agricultural production. In the early 1930s, reductions in planting and in the numbers of draft animals, together with forced requisitions, led to a situation of famine (Nove 1992). Approximately three to five million hunger-related deaths occurred during the 1929 to 1933 collectivisation process in the Soviet Union (Meurs 1999a: 2).

In a tacit admission of compromise and recognition of failures, from 1935 on, collective farm members were able to retain small household plots. Famously, these had much higher levels of productivity than did collectives, reflecting the intensive household labour of women. Still, until the 1960s, agriculture supplied surplus capital for Soviet industrialisation, there was little investment in collective farms and the rural

Figure 4.2 'Woman Tractor Driver', 1932. Photo by Simon Fridland. Acknowledgments to Katardat.org website and Lieven Soete.

population continued to be exploited while collective farm workers and cooperative members remained tied to rural residence through a registration system (Nove 1992; Fitzpatrick 1994). Only after 1965 did government begin to direct more resources into agriculture and to subsidise the sector (Davydova and Franks 2006: 40). Stalinist rural institutions, however, remained in place until the late 1980s (Wegren 1998). In this setting, agriculture changed from being a source of revenue for the state to one that *required* state expense. This was one of the reasons for land reform in newly independent Russia.

Post-Soviet Russia: Initiating Decollectivisation

Land reform in the new Russian state began in 1990 and involved the transfer of state-owned land to individuals and collectives. Subsidies to agriculture were abruptly withdrawn. One of the aims of land reform was to foster efficient, competitive capitalist agriculture based on family farming. The reform did not deliver the results expected. Despite four years of agricultural prosperity, Russian gross agricultural output in 2003 was under 69 per cent of its 1992 level. Food is produced mainly by collectives and (particularly) by household plots. Collectives and household plots combined generate food production, with household plots accounting for 58 per cent of the total agricultural output in 2003 (Davydova and Franks 2006: 41). Private farming, other than on small household plots, has been very slow to develop, reflecting a lack of clarity over individual entitlements (Kalugina, cited in Davydova and Franks 2006), problems with affordability of inputs, and the poor state of rural infrastructure. Thus, a distinctive Russian feature is that even after destatisation, most agricultural land remains collective. Only 10 to 12 per cent of land was individually owned in the late 1990s (Wegren 1998).

A significant question within this context is why large-scale enterprises (LSEs) have persisted. Research by Davydova and Franks in Novosibirsk argues that the breakup of LSEs would entail substantial economic and social costs. Scale economies would be lost by reallocation of large-scale equipment. Many directors of LSEs or ex-collectives also retain concern for their social welfare functions, such as maintenance of the rural infrastructure and provision of employment. Additionally, the authors note that the high level of productivity of household plots relied partly on collectives. They cite a director:

> Today's mass media talks about the development of private farming, that it is a panacea. Well, each family [now] has a lot of livestock. . . . This is only possible because there is a collective farm [ie village LSE]. Without the collective farm, there would be no opportunity to keep livestock in the household. The growth of household production is not

an achievement of rural families alone. It is the collective farm which produces their meat too, because [there] . . . the villagers get hay, grain and piglets . . . (Davydova and Franks 2006: 57).

The authors point out that if the LSEs were wound up or allowed to go bankrupt, some other body would have to pick up social costs. Private farmers do not usually see maintenance of the rural infrastructure or concern for employment as part of their remit.

The ways in which decollectivisation may have had gender implications have not been substantially explored to date, with only a handful of studies addressing the issues (Artemova 2000; Viola 1996; Wegren et al. 2006; but see Wegren et al. 2002). Bridger describes tensions experienced by Russian farm wives. Women's work burdens increased due to the necessity of carrying out labour that on the collective had been mechanised and to the increased burden of domestic labour, reallocated from the collective to the family. Children's labour, too, was important and this often meant their absence from school. Additionally, existing problems of alcoholism and violence escalated (Bridger 1996).

Another account notes that Russian land reform has also affected migration patterns, which in turn impact upon gender relations. In the past (and after administrative controls on migration were relaxed),[6] young people, especially young women, migrated away from the countryside. From 1991, however, rural areas became net recipients of new settlers, many of whom were under 40 years of age (Wegren 1998: 24). The main attraction has been the availability of free or cheap plots of rural land and the collapse in living standards in urban areas with the transition from socialism. As in other contexts, current migrations to the countryside may also be a necessity spurred on by hunger and a quest for food security. And, as elsewhere, subsistence agriculture has become feminised. In parts of Russia, however, this has taken place in the context of reversion to a barter economy (see Burawoy et al. 2000).

COMPARATIVE PATHWAYS TO COLLECTIVISATION AND DECOLLECTIVISATION

Despite the widespread dislike of collectives by many peasants and the destruction it wrought, the Soviet model of agricultural collectivisation was widely followed in its satellite states, in part because the USSR exerted pressure. It was, however, also because of the relative success of Soviet industrialisation and reflected beliefs about economies of scale in agriculture. Many differences existed among socialist countries with respect to the organisation of agriculture, but they also had a number of common features (Nolan 1988). These include ownership of most rural means of production by the state farm or collective, with a large percentage of personal income distributed by collective officials. Cadres controlled labour allocation, and collective farm members were tied to the collective and unable to leave freely. The state, or state organisation, set mandatory production targets. In the case of collectives, any amount produced over the targets would be distributed among members.

Within the common model, a number of different agrarian paths can, however, be traced both during and after the collective period within eastern Europe (Meurs 1999b; Swinnen 2001). In Poland, for instance, there was little collectivisation and most plots were private (Holzner 1995: 13). These usually formed only part of a household's livelihood and were farmed most commonly in conjunction with urban-industrial employment. Croatia, too, retained much private farming. Hungary presents an intermediate case: between 1948 and 1956, a Stalinist model of full collectivisation was followed, but after the 1956 uprising, collectives were allowed to elect their own officials without central interference. In practice, many allowed households to sharecrop collective land (Meurs 1999a: 10). In the Czech Republic, Romania, and Bulgaria, by comparison, agriculture was effectively collectivised. In all cases, small plots were allotted to collective members, with vegetables and fruit usually cultivated for family consumption or for sale. In turn, private plots benefited from goods and services provided by collectives such as seeds, fertilisers, veterinary services, and marketing arrangements.

With the beginnings of *perestroika* (liberalisation) in the 1980s, there was increased emphasis on the need to support private farming, to increase productivity, and to reduce outmigration in both the Soviet Union and elsewhere. With the collapse of state socialist governments and the advent of market economies, agricultural strategies called for decollectivisation, marketisation, and redistribution of land, either to collective farm workers or else to the 'original' owners. A 'European model' of decollectivisation, characterised by land restitution, compensation to pre-revolutionary owners, rapid privatisation, and introduction of marketing rights is thus sometimes invoked (Selden cited in Watts 1998; Swinnen 2001). However, eastern European decollectivisations have varied (Swinnen 2001). In Albania, nearly all land has been redistributed in small plots to ex-collective workers (Swinnen cited in Holzner 2008: 435). In a number of other cases including Russia, decollectivisation has not been thoroughgoing. For instance, elements of collectivisation and cooperatives remain in Hungary and Bulgaria, and Slovakia has retained collectives (Holzner 2008). As Burawoy and Verdery emphasise (1999), responses to post-socialist transitions are not uniform: instead, they may be localised and innovative. In many cases, nevertheless, the picture of post-socialist agriculture continues to be one of large-scale commercially oriented units (either collective or privatised) alongside very small plots. However, a previously viable symbiosis has been replaced by two sectors, often with little connection (Holzner 2008: 436).

Few case studies of gender and decollectivisation exist. Overall, the literature signals continuing neglect of feminist analytical frameworks and assumptions about women's nature as dependants or housewives (Holzner 1995). Those studies which do address decollectivisations have tended to see it as revival of ideologies of domesticity and women's 'place' in eastern Europe and elsewhere (Moghadam 1992; Einhorn 1993). Moghadam has argued that these served to legitimise economic reorganisation and mass unemployment. Soviet-style states had never seriously addressed women's roles and unequal burdens in the domestic sphere, despite some attempts made to campaign for husbands' greater participation in housework in the 1970s and 1980s. Thus, employment outside the home coupled with housework represented a double burden for many women. Additionally, their entry into the formal economic sphere was identified with feminism *tout court*, and feminism came to be linked with the unpopularity of bureaucratic and oppressive state policies. Feminism's relative lack of appeal in eastern Europe has made it easier for ideas of domesticity and essentialised motherhood to re-emerge. Some authors cite a 'patriarchal populist' nationalist retrenchment behind these ideas (Verdery 1990; Holzner 1995; see Burger 2006).

Many accounts acknowledge that collectivisation was by no means completely egalitarian. For instance, the sorts of work that were seen as suitable for men and for women tended to coincide with assumptions that men are more skilled than women, with skilled work usually attracting higher pay, or more work points. Nevertheless, in most cases in Europe and the ex-USSR, women *were* collective members and did gain work points through their labour. This contrasts with Latin America, where there were fewer female members. This made the labour contract visible, and it was individualised rather than being subsumed in family units.

To illustrate comparative approaches to collectivisation and decollectivisation in more depth, I offer two examples of countries that had a high degree of collectivisation: Uzbekistan and Bulgaria. These are followed by case studies of Hungary and Cuba. Whilst within the USSR's sphere of influence, they were distinctive from it.

Uzbekistan

In Uzbekistan, a former Soviet republic, the collective sector still exists, although in weakened forms. Uzbekistan continues to rely on cotton as a main export crop, and the state continues to have strong supporting role (Kandiyoti 2003). Thus, the pace of land reform has been slow. As elsewhere in the former USSR, the agricultural sector has acted as a shock absorber and as a source of livelihood for growing numbers. Within this, gender relations are crucial. Contemporary Uzbekistan presents the 'spectacle' of former teachers and office workers—many of whom are women—vying for plots of land (Kandiyoti 2003). Thus, the country has experienced both reagrarianisation and demonetisation.

Some collectives have been liquidated and the land redistributed. Others have been transformed into joint stock companies. In this new form, the previous *sovkhozy* and *kolkhozy* (now *shirkats*) still occupy the bulk of irrigated land and account for half of the country's cotton and wheat production by value (Kandiyoti 2003: 251). Organisation of production remains unchanged and relies on the family contract system, although production units have become separate accounting units. Collective membership offers both the right to a private plot and to social benefits.

Kandiyoti carried out studies in two such shirkats (joint stock companies) as well as in a liquidated collective or farmers' association. These different patterns of restructuring produced different outcomes for women. In the shirkats, decline in mechanisation has meant that manual labour, especially female labour, and non-monetary exchanges have been substituted on household plots, resulting in an intensification of women's labour. In some cases where population pressure is high, a growing pool of agricultural labourers exists, with women concentrated in the lower echelons. In the ex-collectives, the creation of independent farms has favoured technical or administrative cadres of the ex-collectives who are well placed to obtain rights to new farms; relatively few women, however, achieved such positions. Management of the independent farms is seen as 'masculine' while in small plots and farms, feminisation has taken place. The commercial sector rests on the smallholding sector, itself populated by a largely unremunerated and unrecognised female workforce (Kandiyoti 2003). As elsewhere, diversification of economic activities as a survival strategy has tended to encourage feminisation of agriculture.

These processes have also affected kinship and marriage patterns. Kinship patterns in rural Uzbekistan are similar to patterns elsewhere in Asia and in the Middle East, where patrilineal kinship combines with viri- or patrilocal residence and son preference, despite Soviet encouragement of small and more egalitarian families. The Soviet system, however, restricted any possibility of property transmission. Additionally, it provided a developed social infrastructure with opportunities for off-farm jobs, and women had some access to these. Women also have high literacy rates due to (past) near-universal schooling (Kandiyoti 2003). Agricultural decollectivisation has had impacts on kinship and family as well as on livelihoods. The Soviet marriage contract has been eroded, and *de facto* polygyny has increased. Moreover, the women's committees of the Soviet era are much weakened. Nevertheless, they do continue to provide some recourse in cases of violence. The general context is one of hardening attitudes towards women's work as well as to their mobility and access to public spaces (Kandiyoti 2003: 249). Thus, the decline in women's employment opportunities has been linked with informalisation of the marriage contract and with a decline in women's household power. Women are land hungry, but within a context of deindustrialisation: most women would have preferred more stable and industrial or urban jobs.

Bulgaria

Bulgarian agriculture was very highly collectivised, to the extent that in 1960, collectives accounted for 80 per cent of agricultural land (Meurs 1999a: 14). In the 1970s, the existing collectives were abolished and restructured into huge agro-industrial complexes with acreage up to 30,000 hectares. Although women did not have complete equality within these, the state did provide access to education, medical care, and childcare as well as guaranteeing employment. All of this supported a degree of economic independence for women, including married women. Bulgarian laws providing entitlements for women collective workers in pregnancy and as mothers were particularly favourable (Barbic 1993: 14). Collectivisation formalised women's employment and thus expanded their visibility. Their share of work was subject to remuneration, although often on less favourable terms than men's. In 1967, women constituted 67 per cent of the agricultural labour force (Meurs 1997: 336). However, only a minority were able to access the best paid and higher status jobs such as tractor driving. White collar and some technical positions, however, in this and other cases discussed here, were dominated by women.

Bulgarian decollectivisation was not wholesale. For instance, forms of cooperation exist in order to take advantage of use of farm machinery (Meurs 1997). However, decollectivisation has meant a large-scale loss of jobs. In particular, many skilled jobs have been lost, particularly in white collar work as well as in agriculture. Skilled work is likely to become even more strongly viewed as the province of men. Unusually, Bulgarian women *do* have traditional rights to own land and to organise cooperative work groups, so this may limit their subordination. However, the emerging structure of small private farm units will result in the loss of benefits, loss of women's visibility, and loss of their recognition for work. Due to local peculiarities of compensation for land, few women are likely to become independent farmers as the land will not be in their 'own' villages but the husband's (Meurs 1997: 340). Moreover, collective officials are the people most likely to receive redistributed land parcels.

Hungary

Land ownership in Hungary prior to World War II was dominated by large estates: more than in any other European country (Momsen et al. 2005), and the post-1945 land reform was rather tentative. From 1947 to 1956, collectivisation was encouraged, although it was resisted by many peasants. After the 1956 uprising, government policy combined compromises with peasants and continued pressure for collectivisation. By 1970, collectives included 70 per cent of the agricultural population (Meurs 1999a). However, cooperative members retained rights such as being allowed to work in local groupings and to farm in family units, as well as freedom from state production quotas (Meurs 1999a: 15). These compromises allowed strong elements of patriarchal production into collectives (Asztalos Morrell 1999).

The gender regime which emerged during the collective era was affected by previous kinship and family structures. Historically, the extended patrilocal family was common among the middle and upper layers of the peasantry but poorer peasants and landless agricultural labourers were more likely to live in nuclear families (Asztalos Morrell 1999). On the landed estates prior to World War II, the labour force of permanent manorial workers was male while the seasonal labour force included many women.

In the early period of collectivisation, until 1956, collectivisation brought about a deskilling of autonomous peasant producers, who became more like agricultural labourers within collectives (Asztalos Morrell 1999). Many *gazda* (male household heads) felt themselves to be demasculinised. Inheritance of land and assets of course meant little within the collective, men's decision-making power was curtailed and they lost much autonomy over direction of the household labour process. In the early Stalinist period, women's labour contributions were meant to be evaluated by their capacity to reach the same standard as men despite their unequal burdens. Women were encouraged to enter male occupations such as tractor driving and (outside agriculture) mining. Nevertheless, women's reproductive roles were always seen as paramount (as will be seen).

The post-1956 willingness to compromise with peasant norms included their patriarchal aspects, such as men's domination of management positions, the reinstatement of gender divisions of labour within the collective, and women's continued responsibility for reproductive labour (Asztalos Morrell 1999). While women predominated in some white collar occupations within the collectives, such as administrative work, where they constituted 71 per cent of the total in 1971, they accounted for less than 4 per cent of middle managers and only 6 out of some 6,500 cooperative presidents (Asztalos Morrell 1999: 378).

Asztalos Morrell (1999) has identified three periods within the era of collective agriculture following the 1956 uprising. In the first, the consolidation period of 1956 to 1968, mechanisation of labour was low and so there was need for seasonal workers. Distinctions were made between 'active cooperative members' and 'family helpers', the latter consisting primarily of married women. In this period, the proportion of women in the agricultural labour force was below one-third (Azstalos Morrell 1999: 390). Many women, however, joined collectives to gain entitlement to a household plot for which the precondition was fulfilment of a minimum workday entitlement, set at 100 days for women and 150 for men, annually.

In the second period, from 1968 to 1979, household production became increasingly important. This was also a period of rapid industrialisation in Hungary. From the early 1960s, a discourse of 'suitable' jobs for women re-emerged, exemplified after 1966 by barring of women from a number of occupations such as mining. Women tractor drivers, previously admired, became objects of ridicule. This was in part an expression of explicit opposition from male workers (Asztalos Morell 1999). One indication of the impact of such thinking was that during the 1970s, women's participation in small-scale agricultural work decreased while men's increased. Concomitantly, from 1967, generous maternity and child support systems were introduced, signalling that women's *main* work was in the family, and furthermore, that this was not a 'male' sphere (Haney 1999). Instead of the previous construction of the reproductive sphere as stultifying, it was portrayed as work caring for and maintaining of society (Asztalos Morrell 1999). Women were encouraged either to leave the formal labour force (Haney 1999) or to see waged work as secondary.

In the third period, from 1979 to 1989, labour-intensive production was increasingly switched into household-based production, which was becoming increasingly market oriented. As it was carried out in conjunction with cooperatives, however, it entailed a good deal less risk than would have been the case for free-standing capitalist farmers (Asztalos Morrell 1999: 430). Increased marketisation of agriculture, as elsewhere, signalled an increase in male participation and direction. Asztalos Morrell's interpretation is that this was a way of countering the demasculinisation of collectives (1999). In contrast, women's small-scale agricultural production was carried out as an extension of household duties, and their lower status in the collective also displaced them to private agricultural smallholdings (Répassy 1991). During the 1970s, women's participation in small scale commercial agriculture decreased while that of men increased.

The dynamic stratum of market-oriented agricultural entrepreneurs that emerged with the transition from socialism typically involved men with technical skills. Women participated in some cases, as adjuncts to their husbands (Asztalos Morrell 1999). Whereas in the 1960s most women involved in agriculture were older, by the 1980s more wives were in full-time non-agricultural employment, further encouraging the masculinisation of agriculture. On family farms, women are allowed to work but not to take important decisions. Men were able to resume their roles as breadwinners and heads of household. The 'threat' held by aspects of collectivisation that had an emancipatory gender agenda had partly receded (Répassy 1991).

In addition to changes in the division of labour, an important aspect of this transition has been the privatisation of land, through a mix of restitution; selling of land for compensation bonds; and redistribution of land to employees on state farms and production cooperatives (Burger 2001). Privatisation was nearly complete by 1999, with mechanisms including state farms being turned into limited liability or shareholding companies. Average family farm holdings were 3 acres but the average holdings of companies and cooperative farms were 452 acres (Momsen et al. 2005: 37).

The Hungarian agricultural sector is thus sharply dichotomised. Individual farms constituted 55 per cent of the sector by 2000 (Burger 2001: 263). By 2003, Burger and Szép found that small farms were decreasing in profitability. Women accounted for only 13.5 per cent of small farm managers (Burgerné and Szép 2006: 97).

Hungarian rural areas, as elsewhere, have shouldered much of the impact of economic restructuring since 1990 (Momsen et al. 2005), reflecting the loss of subsidised transport for commuting to cities, the break-up of cooperatives and state farms, and the closure of the industrial enterprises associated with them. Women in particular have become entrapped in the countryside. Moreover, a high percentage of women farmers are elderly, often subsisting on state pensions supplemented with produce from plots. Some rural women, however, have managed to move into new entrepreneurial roles. They have set up as local shopkeepers, restaurant owners, hairdressers, and florists using skills obtained in technical secondary school or on the collective and no longer need to commute to nearby cities. There have also been some opportunities for factory work. Foreign investors, noting the availability of these women, set up factories on green field sites near villages and in some cases even provided subsidised transport to bring in women workers from villages (Momsen et al. 2005).

The Hungarian case illustrates several points. Although within the Soviet sphere of influence, the country was fully collectivised for a relatively brief period. A tradition of male dominance in farming as well as men's resistance to collectivisation played a part in the demise of collectives. Nevertheless, these offered some advantages for women, particularly in terms of employment and subsidised transport and services. Hungary is now a predominantly industrial nation, but its agricultural sector is dichotomised between large- and small-scale farms. As elsewhere, the small-scale rural sector has become one in which women predominate, but they are rarely managers who exercise control over production.

Cuba

Cuba is included because it followed a Soviet model in terms of domination by large state farms and also received much support from the USSR until the early 1990s. The dominance of large state farms was partly due to the pre-revolutionary agrarian structure. In 1946, sugar cane plantations and large cattle haciendas held almost half of all farmland (Deere and Pérez 1999: 64), while only one-third of farmland was worked directly by owners, renters, sharecroppers, or squatters. Thus, most agriculturalists, 60 per cent in 1953, were wage workers rather than peasants (Stubbs 1987: 42). A first agrarian reform after the revolution in 1959 expropriated all holdings over 401 hectares, and a second in 1963 lowered the ceiling of landholding to 67 hectares. Sugar plantations and cattle haciendas were made into state farms which held 71 per cent of agricultural land (Deere

1998: 64). The state farm was seen as the highest form of agriculture, and after 1967, almost all free market transactions including crop sales were prohibited. However, small farmers were not compelled to join the state farms, unlike elsewhere (Deere and Pérez 1999).

Cuba enjoyed highly favourable terms of trade in the Council of Mutual Economic Assistance (CMEA) of socialist 'Eastern bloc' countries. As before the revolution, the country specialised in sugar production. It also pursued a food import substitution strategy. Thus, between 1959 and 1989 its imports of cheese, butter, and milk decreased and the country was able to import wheat rather than flour as its capacity to mill grain had improved (Deere 1998: 66). From 1980, potatoes, peppers, and onions were exported, but beans and rice were imported. Cuban state farms were not highly efficient, and even with a subsidy, the percentage earning profits never exceeded 50 per cent (Meurs 1992).

Consistent with the culture of the region, women were partially excluded from production in the operation of state farms. Women's work on state farms was sometimes unremunerated and was often seasonal and casual (Meurs 1992). Where women were full members, however, they received equal work for equal pay, and increasing numbers of women entered the agricultural labour force from the 1980s (Deere and León 1987: 12). Access to education meant that many professionals such as veterinarians working on state farms were women.

The Cuban agrarian reform, while centralising, did also create a landed peasantry (Deere and Pérez 1999). Tenant sharecroppers were given the right to claim the land they worked, up to 67.1 hectares. By the end of 1963, some 154,000 households held 26 per cent of land. A survey carried out by the University of Havana found that over 68 per cent of peasant households had been constituted as smallholders by the agrarian reform (Deere and Perez 1999: 194–95). A later 1990s survey of smallholders grouped into cooperatives found that 41 per cent of cooperative members had gained land through the initial reform. The mean amount of land held was large, 13.74 hectares, close to the traditional Cuban measurement, the *caballeria* (13.47 ha.).

Under the initial agrarian reform, female-headed households were accorded land titles, although in practice did not always receive them. Even when a woman gained title, a man often took on production responsibilities. In general, where a man was present, he was considered a household head, as elsewhere. To some extent the agrarian reform, *despite* explicit government policies encouraging gender equity, reinforced the notion that women should stay at home (Stubbs 1987). This development accords with the argument that land reforms often encourage housewification (Chapter 3).

In the mid-1960s, peasant producers were integrated into the national planning system, and after 1967, they could only sell crops to the state. They were allowed three hectares for their own provisioning. At the same time, peasants were urged to lease or to sell land to state farms.

Although no direct coercion was used (Deere and Pérez 1999), incentives were employed, since the state farm built modern communities and supplied electricity and water as well as schools, daycare for children, and a health post. Housing was rent free on state farms. The government assumed that the private sector would wither away due to the superiority of state farms, but it soon became obvious that most family farmers had no intention of becoming agricultural waged workers. Much production—of root crops, vegetables, and fruit—still came from small farms (Deere and Pérez 1999: 199).

From 1977, the state began to pay more attention to agricultural production cooperatives (CPAs). On joining, peasants received full compensation for land and the guarantee of basic consumption requirements. Further incentives included preferential access to building material for homes as well as to credit and inputs (Deere 1998). However, if they left the cooperative, they could not withdraw land. The CPAs were farmed and run collectively as autonomous entities, although within the constraints of regional planning (Stubbs 1987: 52). Profits were divided between the state, the coop (for reinvestment), and among members according to their labour contribution. Governance was by an elected committee, and an end of year meeting decided on production, investment, consumption, and profit sharing. The economic viability of Cuban collectives is related to their relatively small size and close social networks (Meurs 1999a: 24).

Cooperatives (called collectives) grew rapidly in the early 1980s. Deere (1986b) researched motivation to join and found women were particularly attracted by cooperatives' potential to reduce domestic labour and the promise of improved housing standards. Peasant wives were guaranteed membership and thus employment and access to their own incomes. Women represented one-quarter of CPA membership in the early 1980s, indicating an interest in economic autonomy from husbands and fathers. A number of women also cited the ease of working collectively as a reason for joining (Deere and Pérez 1999). The development of the CPAs represented the first time that agricultural policy explicitly targeted women and gave them identical rights with men. Their cooperative membership far outstripped that of their state farm participation, where they were only 14 per cent of the workforce. Women formed 12 per cent of cooperative executive committee members in 1985 but 6 per cent of state farm executives (Stubbs and Alvarez 1987: 143). In the late 1980s, women working in agriculture accounted for 10 per cent of females in full-time work; many of these were employed in clerical and technical occupations (Stubbs and Alvarez 1987: 147). In 1984, of 70,000 women employed in agriculture, over 60,000 worked on state farms, and between 5,000 and 7,750 on cooperatives. This was a much higher percentage than elsewhere in Latin America (see Chapter 7). Women appeared to find giving up their land to cooperatives less of a wrench than did men (Stubbs and Alvarez 1987: 148–49).

Tobacco is a crop in which women's labour is relatively visible, and research on tobacco cooperatives and state farms was carried out during the 1980s (Stubbs 1987). Throughout Latin America, both tobacco and coffee are traditionally small farm products whose cultivation depends upon high input of 'family' or women's labour. In Cuba as elsewhere, women's work on small family farms was unremunerated and usually, unrecognised: their participation in decision-making was minimal (Stubbs and Alvarez 1987: 151). This is an example of how women's unremunerated labour in agriculture and small-scale production contributes to 'growth' and to the relative efficiency of peasant units. In state farms in San Luis, women comprised 16 per cent of the full-time workforce and 24 per cent of CPA members. In another cooperative, Cabaiguán, women were 38 per cent of CPA tobacco members (Stubbs 1987: 56). Women also worked in considerable numbers as seasonal or contract labour. In the cooperatives, women's labour was meant to be visible and remunerated. Stubbs found, however, that some cooperatives discriminated against women. In one, 50 per cent of members were recorded as being female, but for a number of reasons including household responsibilities, women did not work regularly or did not wish to do fieldwork and therefore earned only 11 per cent of annual profits (Stubbs 1987: 59). Here, men complained that women often missed committee meetings. Women replied that men should help in the home to allow women time to attend (Stubbs and Alvarez 1987: 155). These women echoed Cuban and other state socialist propaganda attempting to persuade men to participate in housework (Molyneux 2001).

During the 1980s, cooperatives encountered economic problems, but later in the decade their performance improved. Meanwhile a number of independent peasants continued to thrive, usually being better off than cooperative members (Deere and Pérez 1999: 215; Torres et al. 2007). State farm managers recognised that cooperatives had been more successful economically than state farms. Nevertheless, Cuba's agricultural sector, although performing reasonably well in the 1970s and 1980s (Deere 1998), was heavily trade dependent. Thus, the disintegration of the socialist bloc ravaged the economy as the guaranteed market for sugar production collapsed, along with inputs for state farms. (Deere 1998: 62). This ushered in the 'Special Period' of falling production and economic crisis in the 1990s.

Faced with agrarian crisis and mass hunger, the Cuban government took several steps to secure food supplies and basic subsistence. Firstly, emphasis on food production and self-provisioning either in cooperatives or individual plots increased of necessity, given that in the early 1990s, 55 per cent of calories and 50 per cent of proteins were imported (Premat 2003: 87). Stringency and the emphasis on self-provisioning has had particularly important implications for women (Pearson 1997). Measures included the Food Safety Programme (1989), encompassing state investment in irrigation, transportation, and refrigeration (Torres et al. 2007). The government recognised, secondly, that some type of reform of the state farm sector

was needed. Rather than parcelling out land, it formed production cooperatives, called *unidades básicas de producción cooperativas* (UBPCs) In
UBPCs, land remains nationalised but can be leased in permanent usufruct.
Members own what they produce but have to negotiate production plans
with relevant state enterprises. The UBPCs elect their own management
committee and retain profits not needed for investment. UBPCs and CPAs
together farmed over 50 per cent of land in 1998. State farms continued to
hold one-third of land and the private peasant sector, 16 per cent (Figueroa
2003: 526).

Thirdly, from 1994 on farmers' markets were reopened. These have been
highly successful. Farmers' markets have helped the population to meet
basic food needs, providing approximately one-third of Cuba's dietary
needs (Torres et al. 2007: 62). However, it is also the case that the sections
of the population who have access to foreign currency (e.g. through the
tourist industry or through remittances) also have advantaged access to
farmers' markets and to food (Torres et al. 2007).

Fourthly, there has been a partial reorientation away from dangerous
mono-cropping and green revolution technologies. The 'Special Period'
revealed the fragility of high-tech agriculture, and the Cuban government
has attempted to promote organic and reduced-chemical agricultural strategies as an alternative (Levins 2005). The spread of urban agriculture, vegetable production, and organic agriculture is a necessity due to the collapse
of external supports. This highly labour-intensive form of agriculture is
a creative attempt at sustainable self-provision and to seek new markets.
It seems likely that women will play a large role in ecologically aware
agriculture.

The Cuban case is unusual, not least because of the longevity of the collective sector. Women were not as thoroughly integrated into the collective
sector as in other cases examined in this chapter, although their participation in cooperatives was somewhat higher than on state farms. Women's
partial marginalisation fits in with patterns evident in other Latin American
countries (see Chapter 7). As elsewhere internationally, the Cuban collective sector may be giving way to privatised farming. In July, 2008 the new
president, Raúl Castro, made food self-sufficiency a priority. He decreed
that cooperatives and private farmers would be allowed to farm unused
government land (Weissert 2008) in order to increase production.

CONCLUSION

The cases discussed in this chapter indicate some directions for analysis of
gender within collectivisation and decollectivisation processes. In eastern
European and other ex-Soviet cases and in Cuba, women had rights to full
collective membership. This meant that women's work became more visible
and recognised. In practice, not all were able to realise rights. Some women

wished to remain at home as mothers and housewives while perhaps culti-vating small plots, as in Hungary, or else were pressurised into such roles. Part of this pressure was due to male resistance, since full agricultural col-lectivisation also entails some loss of control over the labour and persons of household members. Even in cases such as Cuba, which strongly stressed gender equity, women sometimes struggled for inclusion within coopera-tives. Where women are full collective members, they are still responsible for domestic labour and thus cannot always play a full part as collective or cooperative members. Nevertheless, collectives provided a range of benefits including education and welfare services. In most Soviet contexts, women had the same social security rights as men if they were collective members (Barbic 1993). State socialist governments also provided mass education, and in some cases women were prominent as white collar or administra-tive workers within cooperatives and collectives (Asztalos Morrell 1999; Holzner 2008). Sometimes women entered skilled work, either as techni-cians or in other posts as teachers or veterinarians. The existence of col-lective governing bodies also offered potential for discussion of women's rights within the collective, and sometimes for reporting of gender abuses, as in Uzbekistan (Kandiyoti 2003). Additionally, European collective farm women tended to be better represented than farm women on equivalent governing bodies in private farming communities (Barbic 1993: 15).

In the cases discussed in this chapter, the extent of decollectivisation has been limited or partial, a little-publicised phenomenon. In some, such as Russia, Uzbekistan, and Cuba, many collectives still function, although in very different external circumstances. Large-scale and export-oriented agriculture in collectives and ex-collectives sometimes relies upon small-scale agriculture, as in Uzbekistan. Conversely, the existence of a private agricultural sector may depend upon services provided by collectives, as in Russia (see also Chapter 5). Where opportunities for entrepreneurship exist, a minority of women are likely to be able to avail themselves of these and may find their lives improved. They may welcome the chance to work from home as small entrepreneurs.

The most common patterns in ex-Soviet states, however, include wide-spread female un- or underemployment and feminisation of subsistence farming. With decollectivisation, repeasantisation often takes place, and women are pushed back into informal, unremunerated work. In some European settings, revitalisation of family-based farming with its gender inequities is seen as unattractive, and younger women seek alternative employment (Holzner 2008), leaving older women to farm. In the absence of industrialisation or other economic growth, decollectivisation processes may signal resurgence of traditional gender practices and norms.

The cases discussed in the next chapters are ones in which resurgence of elements of tradition accompany decollectivisation. These may, as in Uzbekistan, also mean a return to kinship- or lineage-based practices. In most cases, these do not bode well for women's security and autonomy.

5 China

From Collectivisation to the Household Responsibility System

China is a distinctive society in many ways—the country's huge size alone lends it immediate importance. Equally important, however, are China's mass social programmes: large-scale collectivisation during the Great Leap Forward (GLF) shifted to the household-based responsibility system and the introduction of marketisation within a relatively short period of time. The Chinese agrarian reform programme also highlights the importance of discussions over the 'optimum' size of agricultural units and of the various benefits of land reform in terms of securing political allegiance, maintaining livelihoods and reinforcing human rights. These policies and shifts in social programmes all have profoundly gendered implications.

China presents an interesting and dramatic study of the intersections of state policy and cultural traditions. The former have often attempted to ameliorate some aspects of discrimination against women, including in landholding and work relations but have often been countered by the strength of kinship and lineage systems. China is notable, too for its recent and rapid industrialisation, which includes elements of rural industrialisation. This further underlines Chinese distinctiveness and has had important impacts on gendered agricultural and landholding patterns.

The chapter is organised as follows. It first outlines Confucian principles of lineage and family organisation, then the background to agrarian reform; the next sections outline the processes of initial land reform and subsequent collectivisation and their gender impacts. Lastly, discussion turns to the post-Mao liberal reforms and rural industrialisation; the chapter attempts to weigh up gains and losses for rural women in the last period. Although I focus on land reforms and land issues, I also give a broad overview of other factors affecting gender relations within and outside collectivisation. An underlying question concerns what factors have most weight in affecting women's social positioning(s): 'traditional' kinship structures, state policies, or market reforms?

CONFUCIANISM, GENDER, AND AGRARIAN SYSTEMS

Lineage and household organisation among the majority (Han)[1] people in China are firmly patrilineal and are usually patrilocal. This is, of course, a common lineage and residence principle, but the Han version was an extreme version of patrilineality. An important point is that marriage was, and is, usually exogamous, so a wife was nearly always from another village, although some regional variation exists (Zhang 2000). This means that wives are outsiders both in the lineage and village (Li, W. 1999). To accentuate this status, a wife usually keeps her father's surname.

Chinese lineage and family organization, and much of society more widely, were based on Confucian principles so that discussion of gender and of household organisation and relations must refer to these. For over ten centuries, Confucianism was the dominant cosmology, political philosophy and doctrine of proper ethics and comportment. Confucianists developed an ideology and system designed to realise a harmonious and hierarchical social order in which everyone knew and assumed their proper stations in family and society (Stacey 1983). The well-ordered family was seen as a basic unit of society and political order as well as being itself a microcosm of society. Confucian ethics emphasise hierarchy and stability in society and within households, with an accordingly strong emphasis upon the welfare of the group rather than individuals within it. Of course, the welfare of some was advantaged over others, along lines of class and generation as well as gender.

The father-son relationship was pivotal to the Confucian system. More broadly, generation plays a large role in Confucian cosmology and philosophy, with older people gaining in status and respect. Because lineages are by definition comprised of both living and dead members, ancestor veneration frequently underpins lineage-based systems of household and territorial organisation. In China and elsewhere, this has implications for how land is viewed and is experienced emotionally, since ancestors' burial sites are placed on lineage land. Thus the Chinese lineage was a corporate entity with its basis in landholding (Freedman 1970, 1971).

Gentry households were most able to support the ideal extended lineage family, but many peasant households, too, consisted of extended families containing three or more generations (Stacey 1983). Confucian ideals, of course, had marked gender implications. Korinek (2003) calls the Confucian patrilineal system one of the most patriarchal in the world, in terms of the exceptionally low status of women, their relative lack of power inside and outside the household, and the constraints on their actions. These emphasised the virtues of filial piety, obedience to elders and to male authority, docility, propriety, chastity, and submissiveness for all women. Chastity for widows was also enjoined. The now-notorious practice of footbinding perhaps best illuminates this ideal of womanhood.

As in other patrilineal systems, life was particularly hard for daughters-in-law, especially while young. Marriages were always arranged and a significant minority of girls came to their husband's home as young children or 'adopted brides', to be trained as daughters-in-law. Regular contact with natal kin was discouraged and in any case was not usually feasible due to constraints of transport.

A variation on the patrilocal residence principle should be noted. In situations in which no male heir existed, a usual practice was to import a son-in-law to continue the lineage of the bride's father. However, this was also a sign of low status (Wolf 1985; Bossen 2002) and uxorilocal marriage tended to be viewed with shame. One element of such feeling may be that it tends to be more beneficial for the status of the woman, who remains with her natal family.

In normal, patrilocal marriages, a daughter-in-law's duty was to assist and to serve her parents-in-law and especially, her mother-in-law. The latter was the main person with whom a new bride would have contact and whose orders would have to be followed on an everyday basis. Relations between a bride and her mother-in-law were famously fraught and were often violent (Wang 1999). Even when husbands were genuinely attached to their wives, the strength of the conjugal bond was seen as secondary to the mother-child bond. A bride's position was highly precarious until the birth of a son and, traditionally, a man had the right to reject and to arbitrarily divorce a wife (Stacey 1983). Confucian families, although ostensibly upholding order, were often sites of gender and generational violence and the bride as outsider was a particular target.

This chapter, of course, discusses the peasantry rather than the gentry. Most accounts (Davin 1976; Croll 1981; Johnson 1983; Stacey 1983; Bossen 2002) indicate that gender relations among ethnic Han peasants did not diverge greatly from the picture presented. Certainly, ideals were shared but practice could be different. In peasant households the husband-wife (horizontal) relation was of more importance than the vertical, lineage relation of corporate lineage groups (Freedman 1970, writing of the southeast). Widow remarriages, although abhorrent to Confucian ethics, were common among peasants, at least when the widow was young (Davin 1976: 77). Peasants valued—or in any case required—strong, healthy women rather than frail creatures (Stacey 1983). The ideal of patriarchal, ordered and self-sufficient households was to have profound repercussions in terms of gender relations in land reform (as will be seen).

Women's Participation in Agriculture

Most studies hold that China was an area of male agriculture (Boserup 1970) in which women's participation was far less than men's, although variation within China exists. Certainly, as in other patrilineal societies, women did not own or inherit land. In some cases, however, widows might become proprietors of land on a temporary basis when they held and worked land for minor sons (Bossen 2002: 353). The Chinese pattern of landholding among smallholders was one of tiny holdings, often fragmented into several segments and of careful and intensive labour in order to grow sufficient yields: Liu calls this the 'gardening' model of Chinese agriculture (2000).

Peasant women cared for children and for in-laws, cooked and prepared food, involving the difficult task of husking rice or millet, fetched water and fuel, cleaned house, raised animals such as chickens and pigs, wove cotton cloth and sewed family clothing as well as making cloth shoes (Davin 1976; Judd 1994; Bossen 2002; Liu 2000). A division often made is one between the wheat-producing areas of the north in which women engaged in little agricultural activity and the rice-producing areas of the south in which they were needed because of the exigencies of cultivating rice (Davin 1976; Johnson 1983). Davin tries to relate women's participation in agriculture to the practice of footbinding—which was somewhat less prevalent in the southeast. Bossen, however, contests this on several grounds, noting that peasant women with bound feet had nevertheless to engage in fieldwork, including wet rice cultivation (Bossen 2002: 105).

Women and their work were considered 'of the inside' while men and their work were 'of the outside' (Jacka 1997), as indicated in the commonly used phrase, 'women weave and men plough' (Liu 2000). However, the gendered division of labour in northern Shangdong was that men grew grain, and women, cotton (Judd, 1994). The 'inside-outside' dichotomy seems to have been an ideal most likely to be achieved in gentry and official households. Thus, as in South Asia, women's participation in fieldwork, particularly in the north of the country, was seen as shameful and only permitted in order to cope with poverty. There were also a number of taboos such as that on women ploughing. Women had a host of capacities for pollution, including of crops and land, especially during their menses (Diamond 1975).

John Buck's surveys of Chinese land use (1957) have been influential (Bossen 2002). He reported virtually no female involvement in agriculture in the north and only slight participation in the south. In contrast, Fei's earlier study of Lu village in the southwest yielded radically different results from the aforementioned (Fei 1939). Women in Lu village were primarily agriculturalists and a major part of the agricultural labour force. They generally did more work than men in the two staple crops (rice and beans) and were active in contributing to subsidiary crops. In a re-study of that research, Bossen confirmed its findings (Bossen 2002). Likewise, Mao Tse-tung, writing from Xunwu (Mao 1960 cited in Bossen 2002), took note of women's heavy contribution to field labour in the 1930s.[2] A reconsideration of women's agricultural roles should emphasise their actual contribution to labour, their roles as agents in their own right, the practices of female labour exchange, and of hiring women agricultural labourers (Bossen 2002). The latter two elements were of importance in the rice transplanting season, which is highly labour intensive. Women were seen as particularly suited for this delicate work, itself linked with nurturing life. Bossen also stresses the complexity of the village economy and of women's roles in vegetable production, raising pigs, and weaving, emphasising the need to view household economies and livelihoods in the past and present in a holistic fashion.

It is difficult to assess the question of peasant women's contribution(s) to agriculture, given the patchiness of data; however, it appears at least possible that it has been thoroughly written out. Contribution to agriculture does not automatically translate into high status for peasant women. However, it seems likely that the status of women in the south is higher than it would have otherwise been, had their contributions been less.

THE CHINESE COMMUNIST PARTY, LAND REFORM AND WOMEN'S STATUS

The Chinese Communist Party (CCP) was committed to improving the status of women, although not to their complete autonomy. Mao himself had been part of the anarchist May 4th urban movement early in the twentieth century, with its radical emphasis on women's liberation (Davin 1976). Before the split between the *Guomingdang* (nationalists) and Communists, both were urban based and committed to reform of practices affecting women. These included footbinding (outlawed in the late 1930s by the Guomingdang), child marriage, forced marriages and bans on widow remarriage. The historic split and subsequent Guomingdang persecution of Communist women and men meant that the CCP had to shift and to rely on its rural and village base much more fundamentally. Subsequently, it would find it difficult to go beyond the gender configurations of rural and peasant villages and households.

The twentieth century was a time of particular crisis in Chinese society and in the Confucian system, with the dissolution of the imperial system, warlordism, and then the Japanese invasion. The crisis reverberated in the rural areas and agrarian crisis was widespread. It was also, and integrally, a 'family crisis' in different ways for elite and peasant families (Stacey 1983). Peasants were often unable to reproduce any family, let alone a proper Confucian one, in conditions of absolute impoverishment, warlordism, and social chaos (Stacey 1983: 67–68).

In the countryside, a number of factors including rack-renting and population growth combined to result in forced land sales (Myers cited in Stacey 1983).[3] Under this pressure, families disintegrated at a rapid rate. Husbands sold wives and daughters much more readily than land (Stacey 1983). Many people, particularly agricultural workers, were unable to marry at all. CCP organisers in Kiangsi found 91 per cent of farm labourers and 70 per cent of poor peasants unable to marry due to poverty (Stacey 1983: 91). Additionally, landlord power over tenants in this period increased their access to peasant wives and daughters. All of this meant both the decline of independent family farming and the decline of peasant men's position as family patriarchs. In this context, land reform was one of the crucial planks of the CCP in this period and later. Land reform would address several issues including peasants' human rights and their political allegiance (see Chapter 1). It would also increase economic productivity—a continuing issue.

In 1927, Mao and Chu The led the remnant bands of their forces, defeated by Guomingdang nationalists, to the Chengkangshan Mountains of rural Kiangsi. They were to 'Devise a successful route to state power in near total isolation from centres of industrial production' (Stacey 1983: 110). For the five years from 1929 to 1934, the Party carried out a land reform campaign in the Jiangxi Soviet, the central Soviet in China. In 1935, the first remnants of the Red Army which had had to evacuate Jiangxi arrived in northern Shaanxi province. While continually at war for two decades, the CCP governed a population of millions during this period (Davin 1976: 21), a time important in testing economic and gender policies.

Both land and marriage laws were passed during the early period of the Soviets. Under the Land Law, all peasants and hired hands were to have rights to land regardless of sex. This radical law aimed to give poor women economic independence. The policy was based on the party line that women would gain liberation through incorporation into the labour force. Marriage laws were passed in 1931 and 1934 (Davin 1976: 28). These closely followed family law in the Soviet Union, giving freedom of marriage choice and of divorce and outlawing polygyny. Registration of marriage and of divorce were required—for the first time removing peasant marriage from a purely domestic and village matter to a more formal and public realm. The extent to which these early marriage laws were implemented is not clear, however (Davin 1986: 30).

Northern Shaanxi was very poor, remote, and mountainous. Its agricultural labour force was male and its gender ideology, conservative—as was noted by the intellectual CCP women who came to Hanan in 1942. Although the party saw participation in production as the key to equality, feminists in contrast said that women were being overworked and expected to play a dual role without recognition. The party's response was that 'full sex equality had been established and that feminism was harmful'. Appropriate slogans, feminists were told, were no longer 'free choice marriage' and 'equality of the sexes' but rather 'save the children' and 'a flourishing family' (Davin 1987: 36–37). Thus, tensions concerning gender issues were in evidence well before 1949.

Land reform was needed to break the power of the landlord class, and even more urgently, to restore a semblance of food security to peasants (Hinton 1998; see Chapter 1). Reform had been carried out in the liberated areas and after 1949 and was the main policy priority in the countryside. In 1950, 75 per cent of the country had not yet carried out land reform (Johnson 1983: 102), but the process was complete by 1952—remarkable, given the size of the population. Peasants were to assign families in their villages into social classes, and struggles against those deemed landlords or kulaks (see Chapter 2) were often violent (Luong and Unger 1998; Lin 1999). The land confiscated from the well or better off was redistributed to the poor and landless. Since average landholdings were under one hectare and usually in fragmented holdings (Lin 1999: 151), 'kulaks' were not in fact wealthy themselves.

The aim was to make private landownership available to the mass of poor, landless, and middle peasants. The peasantry, as Stacey notes, had already taught the CCP the need to avoid alienating the owner-cultivator class: confiscating land from the middle peasantry had led to sabotage during the Soviet period. Thus, the prior experiment had indicated the importance of an alliance with the middle peasantry and its small family farming economy (Stacey 1983: 118). Land reform and the issuing of title deeds elevated the majority of poor peasant households to the middle peasantry and increased production. It also ended the dominance of landlordism and village elites in favour of the Communist Party. By 1954, landholdings of the poor accounted for 47 per cent of the land and that of landlords was down to 2.2 per cent from 40 per cent at the time of the revolution (Lin 1999: 158).

Thus, the Chinese land reform was, and remains, one of the most sweeping to date. It was also highly unusual in that it granted land titles to women as well as men. Young women, particularly among the poor peasantry, were active in land reform campaigns (Diamond 1975; Andors 1976). Women with absent or destitute husbands, widows, and family-less (so-called 'lumpen') women were already forced to make their own way in life and were among the first to join land reform struggles (Johnson 1983: 107). In general, in the liberated areas before 1949, where men were often away in the army, women were at times the main force for land reform (Frenier 1983: 42).

A resolution from the Central Committee in 1948 had instructed land reform cadres to issue title deeds to women even where land was included in family allotments. When this policy was implemented, it was highly symbolic, particularly in such a male-dominated society. Women issued title deeds sometimes heard their own names spoken in public for the first time rather than being referred to in kinship terms—for example, as 'wife of X' (Frenier 1983: 50). Thus, a new sense of individuality was based on material factors.

Women were apparently quick to see the implications of holding land. In an often-quoted passage from William Hinton's *Fanshen* (1966: 397) a woman in Long Bow said 'Before, when we quarrelled, my husband said "Get out of my house!" Now I can say, 'Get out of my house, yourself!'. Another woman cited by Hinton asserted, 'When I get my share, I'll separate from my husband and he won't be able to oppress me any more.'

Despite the 1948 resolution, there were many impediments to women actually claiming title deeds. Firstly, class was assigned to families rather than individuals. Marriage was ideally hypergamous (ie women marry 'up') and so a woman might well marry into a family with a different status to that of her natal family. Assignment of class status was crucial for a household's material situation within the land reform. Women from households deemed to be wealthy also might suffer rape and other sexual crimes as part of class struggle, and this was seen as legitimate targeting. Traditionally, any slur on a woman's sexual honour was a serious accusation. As elsewhere, a traditional antidote was for the woman to commit suicide.

Land was in effect distributed to heads of household. The Crooks' (1979) well-known study of Ten Mile Inn village in Hopeh indicates that cadres routinely turned over a woman's deeds to the male family head (cited in Johnson 1983). Thus, decisions over allocating land to women transmuted into a matter of determining which family received her share of land rights, her natal/father's family or that of her husband. It was mainly widows or divorcees who received their own deeds (Stacey 1983: 130–31). It is likely that only a strong local women's organisation with independent jurisdiction could have accomplished distribution of land to women.

One method of asserting women's land rights might have been through divorce, as the right to divorce had been established by the 1950 Marriage Law. Both land reform and marriage reform were seen as part of an overall democratic social reform by the national leadership. However, the divorce had to be approved by local cadres. They often saw marriage reform as conflicting with the central purpose of land reform: to provide economic security and justice to male peasants (Johnson 1983). In 1949, the CCP's legal committee reported that women had not been allowed to take property with them when divorces were granted in 61 per cent of cases (Stacey 1983: 175). Lower-level cadres often interpreted a woman's divorce petition as a means through which a poor peasant might lose both his wife *and* his property (the latter, through his land rights as well as his stake in her person) and so most petitions by women were refused (Stacey 1983). Wives of peasants continued to be seen largely as family property.

Women asserting land claims were often subject to violence. This was hardly surprising in an atmosphere of ubiquitous violence, and in a Confucian tradition in which violence against subordinates, including women, was a norm. Hundreds of thousands of women lost their lives attempting to assert land rights (Davin 1988: 143).

Thus, radical policies concerning gender and land were not generally implemented or else were partially implemented in a manner not disruptive to household status hierarchies. However, the fact that such rights were granted at all remains of much importance not only historically but in the present. Additionally, struggles for land reform were crucial in bringing the mass of rural women into public and political life, usually for the first time. As Stacey (1982) emphasises, land reform was not a 'restoration' of lineage-based patriarchy. When land, the economic basis of lineage power, was confiscated, this radically democratised and equalised possibilities of a small patriarchy based on family farming. The new family had elements of both the old lineage model and of a newer (in the Chinese rural context) petty bourgeois family (Stacey 1983). Peasant families usually were able to keep and raise the children, including girls, that landlessness had led them to discard. Within smaller families, women's status was enhanced, although by no means equal to men's (Stacey 1983: 134–35). Contrary to the picture painted in Chapter 3, China is therefore an unusual example in which the initial land reform raised peasant women's status and increased their influence. This was, however, from an exceptionally low base.

By 1952, as noted, land reform was complete across the country. An important public campaign targeted the lack of the Marriage Law's implementation, including its provisions on divorce, free marriage, child marriage and ending marriage payments, which were routinely ignored.

COLLECTIVISATION AND THE COLLECTIVE PERIOD

The initial land reform was successful, representing a considerable achievement in terms of lifting many out of absolute poverty into food self-sufficiency. However, land was so thinly spread that it could not provide a basis for prosperity for most rural households (Davin 1976). The government looked to a more collective form to provide a sounder economic basis in the countryside (Davin 1976) as well as to utilise surplus produced for industrial production. Here the arguments discussed previously concerning optimal size of agricultural units are relevant. It was assumed that 'big is best' in agriculture as in industry. Hence, collectivisation was launched soon after land reform had been completed and after (male) peasants had achieved relative security. The collective period lasted from 1952 to 1978 and was then followed by the household responsibility system.

Chinese government strategy was based on that of the USSR and aimed at building the country's capacity to produce capital goods and military materials (Lin 1999). At least 45 per cent of national investment went to heavy industry whereas only 10 per cent went to agriculture, despite the fact that 87 per cent of the population remained rural at the beginning of the period (Lin 1999: 154–55). Wages were low in urban areas but were subsidised by inexpensive housing, clothing, and food as well as by medical care. To secure food for urban rationing, a compulsory procurements policy was imposed in rural areas. Peasants were obliged to sell certain quantities of grain, cotton and edible oils to the state at fixed prices. In order to prevent an influx of rural people attracted by the favourable conditions in cities (and to be fed by low-priced rationed food), a household registration system was implemented to prevent labour mobility. By 1980, the rural population still constituted just over 80 per cent of total population (Lin 1999: 156). Agriculture played an important supporting role in China's heavy industrial development strategy, with agricultural products making up over 60 per cent of foreign exchange earnings in the 1970s (Lin 1999).

The CCP had a fixed belief in economies of scale common to Marxist thought concerning agriculture, despite much counter-evidence (see Chapters 1 and 4). Lin stresses the state's reluctance to invest in agriculture and thereby to improve productivity. Rather, it pursued a strategy in which agriculture would not compete for resources with industry. Moreover, collective agriculture was seen as providing needed mobilisation of labour on projects such as irrigation and flood control at the same time as raising yields in agriculture through application of traditional intensive methods such as closer planting, careful weeding, and use of fertilisers.

Thus, very soon after the land reform had been completed and male peasants had achieved some security, the collectivisation drive was launched—at first, gradually. The first formation of cooperatives of around 30 households was voluntary. In 1953, those in government in favour of rapid collectivisation won the debate in the party, despite the strongly negative example of the Soviet Union. By 1957, 753,000 collective farms, termed 'advanced cooperatives,' with 119 million member households, had been established (Lin 1999: 160). This collectivisation encountered little active resistance.

The size of the collective unit was greatly enlarged in the GLF. The People's Communes introduced in 1958 each consisted of about thirty collectives of 150 or more households with 10,000 acres cultivated. Cultivation of vegetables on very small private plots had been permitted previously but was banned in the People's Communes. As is well-known, the GLF and commune system led not only to agricultural failure but to disaster (Selden 1998; Bossen 2002). Peasants showed their opposition and there ensued a dramatic decline in grain output—over 30 per cent in two years (Lin 1999: 160). This process led to famine, one of the worst in history. Estimates of deaths vary, but they were certainly huge, between 15 and 30 million people (Gray 1990; Aston et al. cited in Lin 1999; see Hinton 1998 for a rebuttal).

After this disaster, the commune system and 'eating from one pot' were abandoned, although collectivisation was not. From 1962 on, agricultural operation and management was delegated to a much smaller unit of 20 to 30 households: a production team. This had effective control over people and land, although land was owned by the state. Team members shared effective use rights to land as well as draft animals, agricultural machinery and storehouses. They had no access to these *except* [my emphasis] as members of production teams (Judd 1994). Labour contribution continued to be calculated in work points and was usually complex, as well as locally variable. Work points, combined with overall production of the collective unit, determined an individual person's remuneration (as will be discussed).

After the GLF, greater attention was paid to provision of modern inputs such as powered irrigation and chemical fertilizers (Lin 1999). Collectivisation, with the exception of the GLF period, did lay a basis for industrialization by facilitating exceedingly high rates of accumulation (Selden 1998: 128). Collectives assured subsistence for most of their members, and food security was one of their most important achievements (Liu 2000; Li and Lavely 2003). So was the survival of their children (Judd 1994), with the crucial exception of the ravages during the GLF. In the collective period, son preference declined, and with it so did female infanticide and abandonment (Li and Lavely 2003). However, this achievement came at a price for the rural population: stagnation in rural incomes, the virtual disappearance of cash transactions, as well as wide rural-urban divides (Selden 1998). Traditional marketplaces were suppressed, despite their importance as centres of rural culture and sociability. Peasants were bound to land through the registration system, discouraging migration to cities. Their incomes were largely paid in kind and pegged at low levels. Thus, some of the impetus for change came from peasants seeking greater autonomy and supplements to their income through trade and handicrafts (Selden 1998). Regional and local inequalities remained, with absolute poverty a problem especially in remote parts of the country; conditions were better in coastal and better-favoured regions.

Gender and Collectivisation

The effects of collectivisation differed, of course, for men and women. In general, this period combined measures that enhanced women's status while retaining strong elements of traditional organisation. Norma Diamond's (1975) work was pioneering in pointing out the continuity of lineage-based relations in the new collective units. Lineage groups often formed the basis of new cooperatives. Of course, these had changed in that the gentry and wealthier peasant leadership no longer existed, and religious functions were strongly downplayed. Nevertheless, within collectives, a group of agnatic kinsmen usually held use rights over land, ponds, forests, orchards, livestock and equipment. Women joined the production team at marriage. Because traditionally—and usually in the present—villages were inhabited by one or two surname groups, a single surname group could usually dominate decision-making processes (Diamond 1975).

Thus, the household often remained embedded in a network of male kinsmen. This situation was strengthened by the Chinese version of collectivisation. Since residence was patrilocal and cooperation was on the basis of neighbourhood proximity, this was more or less inevitable, barring large-scale transfers of population. The lineage basis of agricultural production groups inhibited gender equity. However, Judd notes that although teams did show strong tendencies to surname concentration, the *formal* mechanism of allocation was not kinship but residence. Some conscious efforts were made to counter patrilineal incorporation: when new houses were built in different areas of a village, for example, members would be allocated to a new team (Judd 1994: 26). Despite such efforts, the overall tendency was to support the agnatic basis of organisation and accompanying ideas about gender. The period before the GLF was accompanied by the 'Five Goods' campaign, a conservative pro-family ideology elevating the role of 'socialist housewife' (Andors 1976). After 1955, women who divorced were often not allowed to take their (hypothetical) shares of land out of cooperative units (Diamond 1975).

The Great Leap Forward and Gender Relations

The period of the GLF, however, represented a temporary break not only in agricultural organisation, but also to some extent, in gender relations. The main purpose of the GLF was to attempt a breakthrough in production despite the lack of mechanisation or investment (Selden 1998). Mao hoped to overcome this by radical reorganisation of the labour force, since labour was abundant. Specifically, the 'underutilised' labour of women was targeted (Andors 1976: 35): women were to substitute for men in agricultural production. Thus, the attempt was to involve 300 million women in production. Additionally, the institutional arrangements of the commune altered elements of household work. In 1958, women accounted for 50 per cent of the labour force in agriculture and in some areas where men were heavily involved in industrial production they accounted for 75 to 80 per cent (Andors 1976: 36). Some women, too, were mobilised for large-scale production such as building dams and dikes and in afforestation, working in separate teams from men to meet objections concerning propriety. Women whose household burden was large were encouraged to develop a courtyard economy—that is, engaging in subsidiary production near their homes, usually involving raising pigs and chickens.

The GLF broke the implicit pact with peasant men and began to threaten the household economy (Andors 1976; Stacey 1983). The GLF promoted direct payment of income to individual workers, seriously challenging male control over the household economy (Croll 1985: 29). Furthermore, domestic work was socialised (Andors 1976). Collective services such as mess halls, childcare facilities, and other welfare services that would free women for agriculture were seen as necessary.

During the GLF there was also a brief revival of the politicisation of questions of gender, as had occurred during the marriage law campaign previously in the decade. Generation as well as gender was called into question. The press ran articles with explicit anti-patriarchal justifications; for example, the commitment to communism was to replace traditional commitment to family. The ideological campaign that was part of the cultural revolution also encouraged uxorilocal marriage (Croll 1987a; Judd 1994) as a basis for undermining men's authority within patrilocal families and lineage-based villages.

Hostility to the new arrangements for women's agricultural work was widespread. In particular, male peasants resented the loss of women's domestic labour and erosion of their own authority. Collective dining was a further cause of hostility because of poor planning (Andors 1976: 38). In addition, peasant households relied on fires from cooking to provide heat in winter, so without these, homes were left cold (Bossen 2002). In any case, large mess halls were unable to provide for individual or family tastes in food, and this aspect was disliked by many women as well as men. When economic conditions worsened by the early 1960s, mess halls and other facilities were abandoned, often abruptly. The time of the GLF coincided with other problems: severe droughts and floods as well as sudden withdrawal of Soviet aid added greatly to the result of famine (Selden 1998). However, it is usually agreed that peasant resistance to communes was decisive in their failure. Because overt resistance was virtually impossible in China, most resistance had to be covert, taking forms such as failure to work efficiently or of small-scale sabotage (Zweig 1989). Faced with the widespread decline in production and severe food shortages, communes were effectively abolished; nurseries and other facilities abandoned; and small private family (garden) plots restored. In 1962, decentralisation back to the level of the production team was officially endorsed. Interpreting this transition, Stacey argues that when class struggle threatened to collectivise domestic production itself, and to bypass patriarchal authority over the family and women within it, many male peasants resisted fiercely (Stacey 1983).

Despite the patriarchal fears, women were not treated as equals within the work point system which rewarded work inputs in the collective period, from 1952 to 1978. Land was (and remains) owned by the state but was held by the commune brigade and team. Within this system, the production team of 20 to 30 households was the basic operational and accounting unit. At the end of each year, team income, apart from deductions for taxes and basic needs, were distributed according to the work points each person had accumulated.[4] Work points were sometimes assigned to individuals, sometimes to tasks, or else a combination. Various Sinologists analysing women's remuneration under the work point system find that women were usually paid roughly 20 to 30 per cent less than men (Li 2005). Firstly, as a number of feminists note, women were never remunerated for housework and childcare. Nor were women remunerated at the same rates as men for the same or equivalent work. Some cooperatives gave women seven points for work for which men would have been paid ten points, although in others more effort was made to equalise earnings (Davin 1976: 144).

A number of issues were involved in this unequal allocation. First was the sexual division of labour: women were usually assigned tasks such as weeding, small irrigation projects or subsidiary production, and were accorded anywhere from three to eight work points, but more usually five or six were allocated (Andors 1976: 38) in the 1950s and early 1960s. Even where women did the same job as men, evidence suggested that they received unequal work point allocations (Andors 1976; Davin 1976). Generally, men tended to be classed as 'high quality' labour, and men's jobs were seen as more difficult and thus better remunerated. When work points were awarded according to piece rates, women tended to fare better than men (Li 2005: 289). Another factor was that, as in workplaces elsewhere, women's work records were often patchier than men's due to their withdrawal from work in some stages of pregnancy, postpartum and due to family responsibilities. Additionally, women did not participate in fieldwork during their menses. Meanwhile, men could earn 'big points' on jobs such as irrigation projects during the winter (Li 2005: 296). In contrast, women were sometimes left with no work at all during the slack season (Potter and Potter 1990: 120–21).

Despite inequities, the work point system offered a number of advantages for women and girls. Ironically, this is evident in the very recording of allocation of work points, which made women's work much more evident than it tends to be within households. As elsewhere, women's entry into social or public production meant that they carried a double burden of labour. It was also true that women's work points tended to disappear into the household allocation as the male head of household still usually controlled income. Nevertheless, the fact that work points were awarded to women 'gave face' (Davin 1988) because these were calculated and recorded. The payment of work points was a kind of social as well as material acknowledgement. This practice is likely to have improved women's status, given the harshly patriarchal system. Additionally, because adult women along with their husbands carried the main responsibility for acquiring work points for the household and family, a measure of power shifted from the mother-in-law and father-in-law, who normally would no longer earn the maximum allocation. Changes due to the marriage law and to public campaigns were also relevant. The normal age for women to marry in most provinces was 23, although the legal minimum age under the 1950 Marriage Law was 18 (Gu 1988; Poston and Glover 2004). Thus, young women usually worked for a longer time before they were wed. The figure of the Iron Girl—a female super worker able to labour as men did—was significant in altering the consciousness of rural women. The new cultural value attached to women's work allowed them to see themselves as meaningful persons, although in reality, only a tiny minority of women were Iron Girls (Honig 2000: 102). At the same time, this imagery also indicates that in order to be acknowledged at all, women were enjoined to become like men. Several middle-aged women interviewed in Sandy Island spoke of the collective period as one of 'happiness'. One woman said, 'Now we can eat the same food as our husbands. We can eat wheat flour instead of corn meal and can have meat sometimes.' (Wang 1999: 29). Another said, 'Now we have more say at home.' (Wang 1999: 33).

The revolution also reconfigured village social spaces so that women could appear in public (Hershatter 2002). Schools and meeting grounds became spaces where women could appear outside and still remain respectable. The fields now owned by collectives were spaces in which women were expected to labour with other women. Hershatter (2000) emphasises Shaanxi women's appreciation of the opportunity to meet and to spend

Figure 5.1   Chinese Women on Tractors. Artist: Ding Yu, People's Arts Publishing House, Beijing, 1951. Thanks to Hoover Institution Archives, Stanford University for use of poster image.

time with other young women outside their own households. Davin goes further in writing that young men and women were sometimes able to use collective work and political settings in order to make acquaintance with one another (Davin 1988: 137), although young people did not gain the right to choose their own spouses.

In economic terms, the record of collectives was ambiguous. Selden writes that the period of greatest gain in production *preceded* decollectivisation and was fuelled by the raising of producer prices and increased investment in agriculture (1998: 127). Certainly, successful collectives existed (Judd 1994). However, others emphasise how the continued stress on industrial accumulation meant stagnation in agriculture: China became a net importer of grain from the 1960s into the early 1980s (Lin 1999: 163).

ECONOMIC REFORMS

The era of decollectivisation began in 1978, shortly after the death of Mao and the end of the cultural revolution. The purposes included altering agriculture's position in the economy, increasing rural production, and diversifying agriculture (Croll 1987b). Decollectivisation began with an experiment with household-based production in Anhui and Sichuan provinces, largely due to poor harvests and ensuing hunger (Zhou 1996). It was successful in improving harvests and received support from cadres: by the end of 1982, 70 per cent of production teams had adopted the new system, and by the end of 1983, land had been contracted out to 98 per cent of households (Croll 1987b: 107).

The 'household responsibility system' devolved land and other agricultural resources to households, along with responsibility for delivering a certain amount of agricultural produce. Excess production beyond contracted levels could be retained by the household. Land is not wholly privatised but is held by the state and administrative villages; in this period it was worked on long-term leases, usually of 15 years. The household responsibility system devolved important resources and their management to household level. Contracts are normally signed by the husband as household head. Thus, the landed basis of the collective system was seriously eroded, and an effective repeasantisation occurred. Some argue that the reforms were almost entirely the result of the active agency of Chinese peasants' wish for household-based systems (Zhou 1996). One way this manifested itself in the late 1970s was through bribes to cadres to allow more land for household use. But the reforms gained approval of cadres not only due to low-level corruption but also because of gains in productivity (Zhou 1996: 55).

Decollectivisation was not compulsory, although encouraged. The initial idea was that this would affect mainly poorly performing collectives, often in remote parts of the country. Although most land was contracted out within a few years of the implementation of the household responsibility system (Summerfield 2006), some collectives did persist. Collectives in 1992 accounted for 35 per cent of ploughing and 75 per cent of irrigation (Watts 1998: 181). Li records Hebei villagers' shock at the division of collective property 'It was as if an earthquake were occurring every year' (Li, Z. 1999: 244).

Average family holdings were very small: less than half a hectare, divided between five to ten plots (Summerfield 2006: 139). Land was such an important resource to villagers that government could not risk their dissatisfaction, so the redistribution process was handled carefully and with attention to principles of equity (Luong and Unger 1998). Every individual household member had a land allocation, credited to the household, and allocations were to be readjusted or redistributed every few years to ensure continued equity of landholdings. For instance, when a woman married and moved to her husband's village, her land might be reallocated to a family which had just had a child. The redistributive system gave continued influence to cadres, as well as attempting to ensure that differentiation within villages did not become too great.

Changes in administrative-political levels of organisation were also an aspect of decollectivisation. Communes and brigades were abolished, at least in name, in the early 1980s. The household responsibility system replaced production teams. Brigades and collectives were amalgamated to what became wholly *administrative* units rather than units of production (Judd 1994).

The government failed to think through gender implications of the responsibility system, as the main emphasis was on raising production (Davin 1988). One element might have been the impetus from below for reform, from male and possibly also, from female peasants. Despite the fact that the household was the main focus of attention, the policy was enacted as if gender neutral.

Chinese Development

Although much of the focus of decollectivisation was to increase agricultural production, which indeed was achieved between 1978 and 1984; thereafter productivity in agriculture stagnated (Luong and Unger 1998: 68; Selden 1998). However, the fate of agriculture is only one aspect of the story of decollectivisation. Another aspect concerns the encouragement of rural enterprises: the seeds of these lay in sideline enterprises first developed during the collective period. From the early 1980s, households were permitted to establish enterprises themselves as well as to hire labour (Bossen 2002). Rural marketing systems, suppressed under collectives, were also revived. Rural-based industry, often owned by villages, has profoundly influenced transformation in the countryside and in the country as a whole. Chinese rural dynamism in the 1990s was fuelled mainly by sidelines and rural enterprises rather than agriculture (Watts 1998). The speed with which rural industry sprang up was a surprise to government, which had invested little in it (Zhou 1996). Ownership is split. In 1993, the figures were 43 per cent for state ownership and approximately 37 per cent for collective ownership, 9 per cent for individual and 11 per cent 'other' (Zhou 1996: 236).[5] In 1998, rural township enterprise accounted for 50 per cent of total rural output.

For the last two decades, China has enjoyed very high growth rates. Using cautious estimates, these averaged 6 to 8 per cent annually (Wade 2004: 578), with attendant rises in living standards. Numbers in absolute poverty have dropped, from 490 million people in 1981 to 88 million in 2003 (Howell 2006: 282). Overall incomes are rising, but differentiation has also increased, both within particular regions and between provinces and regions. There is now great contrast between the booming eastern coastal regions, especially the southeast, and the poverty-stricken hinterland in the west. In general, the poorest areas—and households—are those entirely dependent upon agriculture for livelihoods, with the exception of households involved in commercial agriculture (Croll 1987a; Luong and Unger 1998; Judd 1994). A southwestern study by Unger found that an underclass of very poverty-stricken agricultural households were falling further behind, unable to obtain credit for fertiliser and generally ignored by government (cited in Luong and Unger 1998). The brief outline given is by way of a cautionary statement. China's decollectivisation has not resulted in a wholesale 'repeasantisation' due to the industrial boom, especially that in rural areas.

Gender and Liberalisation

Given the country's size, decentralisation, and regional and local differences, it is difficult to generalise about the position of rural and peasant women, as this too has become more differentiated. Entwisle *et al.* (2000) stress the *variety* of household strategies that have occurred with decollectivisation (see also Croll 1987a; Judd 1994), which signals a clear difference from the relative uniformity of the pre-reform period. Nevertheless, some general trends can be identified affecting rural women, especially those most involved in agricultural production.

One trend is the increase in migration, mainly from rural to urban areas. The household registration system was instituted in particular to control urban migration, as well as to keep tabs on the population; registration status is inherited through the mother and is difficult to change. However, migration was never prohibited as such (Yang 2000: 198). Thus, one could move anywhere, but without local registration it was impossible to get work, buy food, or obtain benefits. The growth of the market economy allowed people to circumvent such controls. In response to migration already taking place, government began relaxing controls on urban residence from the late 1980s, allowing rural residents to migrate temporarily, as long as they took care of their own needs for employment and housing. Much of this work is very low paid, temporary, with poor conditions, and no prospects for advancement (Feng 2000; Yang 2000: 199), although it is often better paid than rural work.

Two aspects of this scenario are particularly relevant here. One is the observation that women migrants tend to be young and unmarried. Women often work in factories in order to accumulate for their own dowries (Zhang 2000). A number of rurally registered women are migrants, temporary or otherwise, and this is likely to have important effects on their perspectives and, possibly, their life chances. The second aspect is that male migration as well as growth of men's participation in local industry has resulted in a feminisation of agricultural work, particularly for older women (Jacka 1997; Liu 2000). According to a 1980s Sichuan survey, 60 per cent of all agricultural labour was performed by women. In Henan, 70 per cent of men worked away from their villages (Jacka 1997: 129). Judd's Shangdong study also found a feminisation of agriculture as men took on non-agricultural jobs (Judd 1994). One of the three villages Judd studied was still collective; there, the village head saw agricultural work as of low status, and it was assigned to housewives. In a second, decollectivised, village, mainly older married women undertook agricultural work. A Yunnan study also found that people who were managing fieldwork, gardening and raising pigs were most likely to be older and married (Bossen 2002: 355). Land is still an important livelihood source which most families are loath to give up. Even with rapid industrialisation, an early twenty-first century study found that women's main employment in Hainan continued to be in agriculture (Duncan and Li 2001). However, younger women in Hainan tried to seek lighter and cleaner work where possible.

Although feminisation of agriculture, partly fuelled by male migration, is a significant trend, Jacka identified another, opposite, trend also of significance: the *withdrawal* of women from agriculture in areas with surplus labour and little alternative employment (Jacka 1997). This trend is related also to mechanisation of agriculture after decollectivisation, since 'excess' labour had previously been absorbed in social projects such as irrigation works. Particularly in remote and underdeveloped rural areas, little alternative employment may be available, and some peasant families have withdrawn women from the fields so that they can be available for domestic work (Jacka 1997: 136). This scenario does appear to signal a classic repeasantisation but is unlikely to constitute the main trend.

Domestic ideology has also been deployed flexibly—or perhaps, cynically (Jacka 1997). Where agriculture has become feminised, the notion of 'inside' has been reconfigured: agriculture is (re)defined as inside work although customarily it was 'outside' and man's work. Traditionally, agricultural labour was defined as too heavy for women but now it is more likely to be seen as a natural extension of the domestic role (Jacka 1997: 136).

In addressing the gendered complications of decollectivisation, some of its aspects and those of market reforms and attendant, wider social change may be seen as 'positive' for women. Such a generalisation immediately raises others, of course: which women? in which class? regional and ethnic positionings? at which stage in their life cycles? The fact that household livelihood strategies are now highly variable (Entwisle et al. 2000), combined with the size of China's land mass and population, adds to difficulty in positing trends.

The first and perhaps most important positive aspect for most women, is that standards of living have risen across most of the country (Muldaυin 1998: 101; Summerfield 2006). For instance, in a remote Shaanxi village children no longer wear ragged clothing (Liu 2000: 113). In Lu village, Yunnan, cash incomes had risen considerably, even taking inflation into account (Bossen 2002: 196). One woman said, 'even a landlord of the past does not live as well as his son today' (Bossen 2002: 219). All villagers have access to electricity and many have televisions, bicycles, and running water. As elsewhere, women are highly unlikely to benefit in equal measure to men but nevertheless usually share in increased income.

Housing is an important indicator of living standards (Li, W. 1999). Unlike in urban areas, houses were always privately owned in the countryside. A provincial survey in Hubei (in central China) reported that average numbers of family members were falling and space for each individual had increased (Li, W. 1999: 235). Various accounts (see Liu 2000: 85–86) note the spate of house building and house improvements in rural areas. However, again women do not benefit equally. Widows' legal inheritance rights to the shared house were rarely enforced (Li, W. 1999).

A second benefit for many women relates to changes in their economic roles. This is evident particularly in roles they assume in managing agriculture on family farms, in sideline production and in rural enterprises (Zhou 1996). As mass phenomena affecting very large numbers of rural women, these are relatively new developments and are significant. Women's assumption of such roles, as elsewhere, usually comes at a price, particularly that of increased work burdens. Without assuming that economic roles translate automatically into the ability to exercise authority, women's economic importance is likely to increase their influence within households. Women's roles in agriculture have had positive impacts on their ability to make decisions, as well as in terms of public 'face' (Li, Z. 1999: 251).

The development of sideline production has had a significant impact for rural women. In Shandong, village women were able to work in local factories even after marriage. In this example, the factories were weaving and dyeing enterprises, traditional female activities (Judd 1994). However, in other cases married women were not expected to work in factories. Household sidelines can be an attractive alternative. Goods produced and sold range from the traditional (e.g. chickens and noodle bars) to assembling goods such as radios for export (Summerfield 2006). Studies elsewhere have also indicated that petty commodity production is an economic niche that can provide survival and some autonomy for rural women (see Davison 1993 on Malawi). In sideline production, women are better able to control their own conditions of work. They also retain all profits after necessary and capital expenditures. Thus, sideline production was a practical strategy spontaneously devised by rural women and then later advocated by the Women's Federation as legitimate (Judd 1994: 237). Sideline production is particularly convenient for older women, given their ties to the household. Zhou argues that for most women, it is easier to deal with husbands or fathers than with unrelated men outside the household (1996). Patriarchal controls faced by rural Chinese women exist not only inside the home, but also in concentrated forms *outside* the household (Judd 1994). Within this difficult context, women may be best off working for themselves, and sideline production provides the nearest possible equivalent.

A third set of developments concern the size and forms of household and, relatedly, authority structures within them. Population policies were first introduced in 1970, spurred on by population growth, especially in rural areas. Resolution of the agrarian crisis post-revolution had enabled families to reach the Confucian ideal of having several sons (Momsen 2004). From 1978, urban families were allowed to have only one child, and rural people and ethnic minorities, two. By 1992, the policy resulted in fertility levels below replacement rates (Momsen 2004: 56). The sex ratio is also highly skewed, resulting from son preference and the near absence of welfare systems in rural areas. Official figures in 2001 indicated that there were 120 boys for every 100 girls (*People's Daily Online* 2004a). For women, the population policy has had two effects. One is that less labour has to be expended on childcare. Women's reduced childcare burdens means that they are more available for other types of work. However, population control has had many negative aspects, including increased surveillance of women—indeed, of their bodies—as reproductive beings (Bossen 2002; Liu 2000). In reducing the number of children in a household, the 'two-child policy' has diminished the size of the whole unit and rendered a Confucian-type vision of a large, complex extended family less feasible. In the longer term, reduced population size may benefit women as well as men. However, the population imbalance itself is a sign of women and girls' continued low status.

A related trend is for an increase in the percentage of nuclear families in the countryside. A study of five provinces in the early 1990s indicates that 65 per cent of rural couples lived only with the husbands' parents. The percentage of nuclear families was growing (Li, W. 1999). In Wang's small Sandy Island study (1999), few women enter extended family settings at marriage, and these situations were seen as temporary. Although the southeast is likely to differ from the rest of the country, it might nevertheless indicate a trend that is emerging elsewhere. In Lu village in the southwest, for example, extended families were rare and plural marriages had disappeared (Judd 1994: 189).

Larger numbers of couples marry *within* villages and mixed surname villages have become more common (Zhang 2003: 269). Nevertheless, a degree of shame concerning uxorilocality still exists (Bossen 2002). Heather Zhang writes of more practical obstacles. In the past, there was strong resistance to campaigns for daughters' inheritance rights and for uxorilocality, especially from agnatic kin who stood to lose inheritances. Uxorilocal marriage had become more acceptable in the north in the early twenty-first century (Zhang 2003). Local cadres tended to see employed married women's petitions to remain in their natal villages as *competition* by women for scarce land resources (Zhang 2003: 266). As noted, patrilocality or virilocality are some of the most important underpinnings of village patriarchy, so trends to uxorilocality are likely to increase wives' room for manoeuvre.

Changing family forms have affected authority relations between children, parents and in-laws. Fathers have weakening authority over sons, for instance, as a result of previous household fission (Judd 1994: 188). Most attention is focussed, however, on the hierarchical relation between mothers- and daughters-in-law. This still exists, but the nature of this relationship differs within nuclear family settings. Daughters-in-law are still expected to help their mothers-in-law with work (Bossen 2002), but relations now are more reciprocal, and mothers-in-law may also be asked to assist. The main help given is with childcare, but occasionally assistance with other work—for instance at planting or transplanting time—may be requested (Wang 1999: 32). However, daughters-in-law are still judged to a great extent on whether they can maintain amicable relations with their mothers-in-law (Jacka 1997: 61).

Countering the positive developments, some recent trends in landholding and political representation are negative for women's power and autonomy. It is important to recall that with the partial exception of the GLF, Chinese collective agricultural units were usually based on a patrilineal principle (Diamond 1975; Stacey 1983). Land groups (three-eight households) still tend to be kinship based (Judd 1994: 50). Elements of popular religion emphasising ancestor veneration or worship are undergoing a revival (Li and Lavely 2003). Part of the contemporary basis of patriliny rests with formal administrative structures based on territorial units. However, the micro-dynamics of village social relations usually 'escape the discourse of patriliny' (Judd 1994: 57). Similarly, Liu argues emphatically that, although clans still exist, the *corporate* character of the lineage has disappeared (2000: 79). Decollectivisation has enabled land to be held by a wider range of social units than lineages *per se*. In effect, however, some lineages do hold land and resources (Judd 1994).[6] Households themselves have increased in importance. The term 'androcentricity' is more appropriate than 'patrilineality' for contemporary rural China (Judd 1994). Androcentricity nevertheless remains male domination and is evident in structures and practices of village governments, informal male-centred land groups, and in household organisation.

A note on political representation and village government is apposite here. Although male domination of the political sphere in China, including in the Party, has a long history, the CCP did make efforts to recruit female cadres (see e.g. Hershatter 2000, 2002; Wang 1999). The particular combination of repression of civil society movements and renewed household and village-based androcentricity in China is unlikely to be favourable for women's political representation. In general, women's representation at higher political levels declined dramatically in the 1980s and 1990s (Judd 1994; Howell 2002), although in 2001, 21 per cent of Deputies in the National People's Congress were women (Howell 2002: 43). However, in the same year there was not a single woman in the Politburo, and only three women have ever been members (Howell 2002). The Party has always been a male preserve and at local levels, women find it harder than men to become party members. This too is affected by marriage residence; membership implies a long period of scrutiny and so marrying-in wives are disqualified in the period in their life cycles when they would be most likely to join (Judd 1994). Still, Bossen found that, with all its faults, the local Party in Lu village managed to maintain some political presence for women as team leaders, family planning representatives and even as agricultural technicians. Nevertheless, in this example women's representation on the village council declined from 1990 to 1998, from 17 to 11 (2002: 358). The *China Daily* reported the comment of a United Nations Development Fund for Women (UNIFEM) official that although rural women had become the backbone of agricultural production and family life in rural areas, this had not brought about an automatic increase in their political participation. 'Women's voices are largely unheard or rejected outright when it comes to decision-making in rural China' according to Lin (2004). Relevant to this point is the pattern found by Judd, that it was acceptable for women to exercise power only over other women, not over men; discussion of gender conflicts was avoided and women exhibited public deference to men. Nevertheless, a few women have become village heads (Lin 2004), so some change may be taking place.

A further detrimental aspect noted with decollectivisation was the withdrawal of many girls from schooling (Davin 1988). Decollectivisation increased the need for child labour at a time of population restriction. Lack of a welfare net or adequate medical intervention increases women's burdens of work, as did the withdrawal of the collectives' services: A crisis exists in provision of rural health services (Howell 2006). The withdrawal of services as government retreats is also linked with the sex ratio imbalance. A real need for care of the elderly exists, and traditionally this responsibility falls upon sons. As recognised by government, lack of provision of welfare and for retirement has fuelled sex-selective abortions and neglect of female infants. Withdrawal of the state also has infrastructural and environmental impacts: women's (and men's) work burdens are reduced in that it is no longer compulsory to contribute to collective projects such as building dykes or maintaining irrigation systems. Without similar, substitute mobilisations at village level, infrastructure may deteriorate (Muldavin 1998: 103)

Decollectivisation and an attendant decline in the power of cadres were welcomed by many. The loosening of controls over men's power within households as well as a need for labour, however, has been accompanied by lowering of the average age of marriage to the current legal minimum of twenty years for women (Judd 1994: 229; Zhou 1996: 183). Zhang found that the age of marriage for men was also declining quickly (2000: 66), and many marriages for both sexes were well below the legal minimum age. Given that marriages in rural areas are normally arranged, this change is unlikely to signal empowerment. There has been a reversion to the practice of unregistered marriage in Shanxi province (Zhou 1996), a pattern also found in Uzbekistan by Kandiyoti (2003). Without registration, women's legal rights are compromised.

Women do have formal inheritance and economic rights which were strengthened by the 1980 Marriage Law and then the Women's Law of 1992 (Summerfield 2006). The Women's Law of 1992 was passed in reaction to growing discrimination in hiring and firing with liberalisation, and it mentions women's rights to housing and to land. The updated Marriage Law in 2001 notes the rights of divorced women to hold land (Summerfield 2006). However, legal entitlements are countered by various levels of public and private patriarchy partly reflecting men's strengthened household power. Men's ability to act with little reference to their wives and children is facilitated by lack of effective political and legal recourse for women at household, local, and higher levels.

Changes in Land Allocation

Changes in the law on landholding have tended to weaken women's legal rights in a situation of growing pressure for land. In 1998, a Land Administration Law was adopted, increasing the term of household use rights to land from fifteen to thirty years. The law exerted strong pressures against the periodic reallocations or readjustments to ensure equity that had occurred for the previous twenty years. The 1998 law was enacted to support long-term investment in land and as previously was enacted in a gender-blind fashion. The lengthened period of adjustment in village landholdings has been highly detrimental for women and has increased female landlessness, which is particularly pronounced among women who have married and moved to their husbands' village (Li and Bruce 2005: 320). One survey of 1,200 households and sixty villages in six provinces (Linxiu cited in Li and Bruce 2005: 316) found that 60 per cent of women had to wait for reallocations to receive land, and 24 per cent were without shares in land. Where women lose their land shares, they risk not only loss of bargaining power but total disenfranchisement, as land allocation is key to being a full citizen of a village. In late October 2008, further changes were passed by the

Party's central committee. Land leases were increased to seventy years, and plots can be used as collateral for loans; these changes will almost certainly have further implications for women's land rights (Hong 2008).

A key element in women's weakened land rights is the increased autonomy given to villages, which allowed for renewal of male dominant cultural values, suppressed to an extent during the collective period. Additionally, lack of clear instruction from government greatly exacerbates this situation. A feature of Chinese legislation is often its vagueness, leaving much scope to local interpretations (Howell 2006). I argue that local interpretations tend to be negative for women's rights. (See also Beall 2005.)

The Women's Federation has voiced concern about land rights lost by women, and some of its recommendations were incorporated into a new Rural Land Contracting Law in 2003. Article 54 (7) creates civil liability in cases when the civil contracting party violates women's land rights. Additionally , it strengthens legal provisions for widows, who already had land allocated in their husband's village. It also protects women marrying into another village from reallocation of their land within the natal village (Summerfield 2006), although no protection exists against the actions of the woman's natal family. Women are often reluctant to claim inheritance rights in their natal households (Zhang 2003). As elsewhere, this is due in part to their need for a fallback position and reluctance to precipitate disputes with their brothers.

Despite legal provisions, it can be difficult for women to obtain effective remedies, whether in the family or at village levels. In many cases, complaints to township and village offices or to provincial government were ignored. In others, women have won legal cases, but village leaders bluntly refused to implement the judgement or to return the land. In Anhui province, 45 married women in one village had land taken away forcibly by the village committee. They went to court and won, but the village leader said, 'You may have won the suit but we are not going to give you anything.' (Li and Bruce 2005: 319–20). This scenario indicates that any legal pronouncements about women's land rights may be hollow unless they can be enforced. As this book argues, the issue is not confined to China. Given decentralisation and the partial withdrawal of the state, such enforcement may become less likely. However, some provinces are acting to protect women's land rights by providing detailed guidance including in cases of divorce and for different types of marriage residence. The All-China Women's Federation campaigns proactively on gendered land issues (Li and Bruce 2005).

Issues of women's land tenure rights operate against a fraught background. In recent years, land issues in China have become prominent due to a series of rural riots or uprisings. In 2004, the government admitted 74,000 'mass incidents' involving 3.6 million people (Watts 2005, 2006); most of these incidents were rural and centred on land ownership rights, corruption of local government officials, and environmental issues. Land is often being sold off for development in an environment in which the gap between the incomes of rural and urban people is increasing. In a general context of erosion of popular land rights, growing pressure on village land and *de facto* privatisation, the prospects for women's rights in land are not encouraging.

CONCLUSION

The Chinese case dramatically highlights the conundrum of gender and the peasant question. Following a successful initial land reform which gave a measure of food security to many and which consolidated peasant patriarchy, collectivisation ensued. Affecting the mass of the rural population, collectivisation hugely enlarged the size of agricultural units and socialised much of the work within these. The GLF attempted to socialise even some domestic labour. Famously, it resulted in disaster. A mixture of drought, misplaced central directives, and peasant withdrawal from production resulted in one of the most widespread famines in human history; opposition to loss of control over women forms one aspect of this story. In this case, agricultural policy placed social transformation based on ideas about the optimal size of farm units above food security.

In the modified collectives that followed, household heads were not in a position to assign tasks beyond the household; their decisions were largely confined to those concerning consumption (Davin 1976; Croll 1987a). Thus, collectivisation reduced gender differentials by taking decisions about work allocation outside the home and into a more public sphere. In China, this sphere overlapped to a considerable extent with kinship and lineage structures (Judd 1994; Stacey 1983).

With liberalisation from the late 1970s, raising agricultural productivity was a central concern. The household responsibility system has had mixed results for women, as far more control returned to household heads. Women have more discretion over prioritisation of work tasks which previously were allocated by cadres; this lack of control was often disliked and resented. However, they are now (again) subject to the will of the husband rather than a more collective or potentially public sphere. State-directed population policies, too, have had a strong impact: increases in skewed sex ratios at birth indicate continuing strong son preference. This is perhaps not surprising given the continued strength of the patrilineal principle, the need for welfare in old age in rural areas, and the fact that there is little opportunity for a thoroughgoing gender critique.

Within agriculture, the predominant pattern now is feminisation of agriculture, especially for older married women, with men usually engaging in other occupations. Feminisation occurs because agriculture is less favoured and lower status work; nevertheless, it seems likely that doing farm work disrupts the traditional status of women as 'inside people'.

The long land leases granted in 2003 are likely to be important steps on a road to privatisation, although at present, ultimate title is held by the state. In theory, women should be able to claim their shares of land allocated, but a number of testimonies indicate that women find it difficult to claim land in practice. Patrilocality is often an impediment as women's rights are not transferred upon marriage and may be completely denied upon divorce. Land rights in most cases are still conceived as household rights, under male control. When women head households, they do benefit. And in some cases good practice of listing both wives and husbands on registration deeds exists. Making such practice mandatory by addition of transparent joint titles requiring a second (ie wife's) signature would be an important step to equity (Summerfield 2006: 155).

Another important aspect of the story is China's rapid industrialisation, including rural industry. China is highly unusual in the success and wide spread of rural industrialisation. It is worth stressing that this reflects not only market factors but also the intervention of the state as well as other collective bodies, including local government and village committees. Decollectivisation of agriculture has been accompanied by development of rural and 'courtyard' industries. Thus, this second redistribution has not resulted in a classic 'repeasantisation' except in more remote areas.

Opportunities afforded through industrialisation and entrepreneurship will be affected by social class factors. Some women belonging to families with more resources, stronger social networks and those with entrepreneurial inclinations may be advantaged. Thus, differentiation will have impacts for women as well as men in rural areas, and it is possible that a class of rural women privileged in their own right will emerge.

The discussion of China's trajectory has stressed tension between views emphasising strengthening of the rural patriarchal or androcentric family with decollectivisation and those emphasising opportunities arising through the advent of a market economy. Countervailing trends exist, but it appears likely that a large group of rural women will be left behind working on family farms but with little ability to control production and declining access to land.

6   Viet Nam
Egalitarian Land Reform

The Vietnamese case of agrarian reform reflects its twentieth century history of colonialism and of war. The impacts of the country's division and subsequent reunification are evident in regional differences in agricultural systems. Viet Nam remains overwhelmingly a rural nation, with over 74 per cent of the population in the countryside (UN Population Division cited in Food and Agriculture Organisation [FAO] 2005: 67). Thus, agriculture is far more significant in the economy than in many of the cases discussed so far.

This chapter begins by briefly outlining the history of land and agrarian reforms in north and south. It then examines the nature of the dominant gender regime in Viet Nam, as well as legal changes introduced by the Communist Party. The bulk of the chapter discusses collectivisation and then liberalisation policies, known as *doi moi* (renovation). Discussion of these matters is centred on the north, as it was more fully collectivised than was the south. Collectivisation in the north took place in the context of wider changes in women's social position. The decollectivisation and doi moi policies made another set of sweeping changes, restructuring agricultural units at household and wider levels. Viet Nam at this time enacted an exceptionally egalitarian agrarian reform. This allows examination of the effects of land redistribution with decollectivisation in a context that remains largely agricultural. Thus, Viet Nam is an important case study.

AGRICULTURE AND AGRARIAN REFORMS

Traditionally, three types of tenure obtained in Viet Nam: state, communal (ie the land of the village or the pagoda) and private, constituting 80 per cent of land (Tran 1999: 96). Land owned by local people was further categorised into large scale (holdings of over 3.6 hectares), medium (between 1.8 and 3.6 hectares), and small (holdings of under 1.8 hectares). In the early nineteenth century, 92 per cent of northern households had holdings categorised as 'small'. In the south, with a smaller population and large amounts of reclaimed land, larger holdings were common (Tran 1999: 97). Throughout the country, however, the great majority of peasants cultivated the land of landlords. Some women cultivated land, acquiring this mainly through inheritance of family land.

The north of the country experienced far more complete collectivisation and for a much longer period in the twentieth century. Agrarian reform was instituted beginning in the late 1950s in areas under Viet Minh control, followed shortly after by full collectivisation, complete by 1968. Thus, collectives existed for over two decades. Collectivisation appears to have been carried out without force (Deere 1986a), perhaps aided by the very small size of landholdings as well as solidary traditions. It resulted in an egalitarian redistribution of land. Subsequently, in the early 1980s, a new approach known as doi moi brought in policies of liberalisation. Both these policies demonstrate gains and losses for women, documented in a number of ethnographies of rural women (e.g. Lilejström et al. 1998; Bich 1999; Gammeltoft 1999; Korinek 2003).

In the southern areas, landlord control and inequitable land redistribution were a strong component of nationalist and socialist struggles in the twentieth century. French colonial rule inadvertently strengthened landlord control (Prosterman and Riedinger 1987). Though a limited land reform did take place in the 1950s, various delays and a counter-reform meant that by the end of 1967, 275,000 hectares had been redistributed to approximately 130,000 households on a 'household' or private basis (Prosterman and Riedinger 1987: 118). Thus, only one-tenth of tenants and the rural poor (Prosterman and Riedinger 1987: 126) had benefited. From 1970, under Nguyen van Thieu and in the midst of war, a more extensive reform redistributed 44 per cent of agricultural land to 75 per cent of tenant farmers (Prosterman and Riedinger 1987: 139). Some households joined agricultural cooperatives and a minority chose to collectivise. However, villagers could retain the means of agricultural production such as water pumps and small tractors, and household services were often contracted to cooperatives. Collectivisation, therefore, was always partial (Luong and Unger 1998: 62).

The victory of the Democratic Republic of Viet Nam (DRV) in the long US war and reunification of the country in 1975 saw expansion of collective production, central planning, and a redistributive economy to the south and its intensification in the north. As elsewhere, the government followed orthodox Marxist strictures in assuming that large-scale agriculture was more efficient than small farms and that collectives were more politically acceptable than private agriculture. Labour was organised into industrial and agricultural production cooperatives to meet production targets so as to make compulsory deliveries of grain to the state (Korinek 2003). Workers were organised into brigades and paid in work points. Nevertheless, there were cracks in state planning: local or provincial authorities often managed enterprises and an informal economy flourished in the south after reunification (Korinek 2003: 59).

GENDER IN VIET NAM

Traditional Gender Regimes

Gender regimes in Viet Nam bear some resemblance to Han China, though there are differences. The discussion here concerns mainly the majority ethnic group in Viet Nam, the *kính*, but a large number of ethnic minorities exist in the country. Most of the country is Buddhist, but minorities follow traditional religions or Roman Catholicism.

Ten centuries of imperial colonial rule from China meant that a Confucian ethic prevailed. This included patrilineality and son preference. The 'three obediences' for women (ie to father, husband, and adult sons) were also enjoined. Kinship patterns, however, were, and are, also strongly influenced by the more egalitarian gender patterns of southeast as opposed to east Asia (Bich 1999; Kabeer and Tran 2002). Several factors indicate that Vietnamese rural women enjoyed greater autonomy than did Chinese village women. Firstly, footbinding is not mentioned in literature on Viet Nam, so this painful and highly dangerous practice limiting women's mobility was apparently absent. Secondly—and perhaps relatedly—women were economically active and were recognised as being so. This applied particularly to women's positions as traders in most peasant households and even in upper-class families (Bich 1999; Boserup 1970). The activity gave them physical mobility, increased status, and accompanying recognition that women could handle money, thus they often controlled the family 'purse', especially in peasant families (Bich 1999: 39). Thirdly, both sexes have long been active in agriculture and the gender division of labour was relatively flexible. Men usually ploughed but could not farm without women to perform tasks such as transplanting rice seedlings (White 1982). Other tasks such as harvesting were less gender-stereotyped although transplanting rice seedlings is seen as exclusively female (Korinek 2003: 76). Fourthly, although Viet Nam is patrilineal and son preference exists, a married woman retains much stronger links to her natal family than in most of China, indicating a strong bilateral tendency in kinship (Gammeltoft 1999). Women could also customarily inherit family property. Early nineteenth-century records for several villages show that up to 24 per cent of landowners were female, although they usually held very small amounts of land, under five *mu*[1] (Tran 1999). Fifthly, residence norms made it possible for wives to retain links with maternal kin. Residence in Viet Nam was patrilocal or virilocal, but most Vietnamese villages are endogamous rather than exogamous as in China (Bich 1999; Luong 1989). This creates a very different scenario for women who remain in their natal villages. Nevertheless, women's rights to houses and land are much less strong than men's. Overall, women in Viet Nam had a higher status and stronger bargaining positions than women in China. Yet they did remain subordinate to husbands and (especially) to in-laws. Bich (1999) notes that within houses, the sexes traditionally had separate locations. The kitchen was 'women's' and seen as of low status. In other respects as well, family life followed Confucian mores and practices. Marriages were accompanied by bridewealth and dowries; child marriage was frequent; and polygyny and concubinage were permitted, particularly in upper-class families. Husbands' families usually had custody of children in case of divorce. As in China, violence against wives, particularly younger wives, was normal.

Communism and Gender

The Vietnamese Communist Party was founded in 1930, presided over by Ho Chi Minh. Shortly thereafter, the Communist International directed a change of name, to the Indochinese Communist Party. The Vietnamese Women's Union (WU) is the party organ dealing with women's issues. It has existed continually since 1930, when women in the north were mobilised in anti-colonial struggles. As elsewhere in the ex-socialist world, civil society organisations either do not exist or are discouraged or repressed. Given that no grassroots women's organisations have been permitted, the WU is the main body representing women. It is particularly important in rural areas, where 80 per cent of women live (UNDP 2001). Although a Party organisation, it has a degree of autonomy (Eisen 1984).

As elsewhere in the socialist world, marriage reforms were enacted, aiming both to outlaw 'feudal' practices and to bring women into the formal workforce. Viet Nam was divided into north and south in 1954. In the north, the 1960 Marriage Law outlawed polygyny, concubinage, child marriage, and forced marriage; banned dowry and bridewealth; and set minimum marriage ages for men at twenty and women at eighteen years (Bich 1999: 58). These measures were 'standard' socialist marriage laws. Although these were very hard to enforce, the Vietnamese law went further, giving women more rights to child custody, legitimising children born out of wedlock, outlawing wife beating, giving wives rights to communal property, and banning the exercise of power by one spouse over the other. Thus, the law eliminated most articles of the *Gia Long* code, based on Confucian principles (Bich 1999). It also contrasted with the marriage law of the south, which made the husband the legal head of family and which required that divorce be authorised by the state (Bich 1999: 59). Later, in 1986, this law was strengthened by another marriage law.

COLLECTIVISATION IN THE NORTH

In the north, the Viet Minh gave women a share of land in areas they controlled from the 1950s (Tétreault 1994: 114). Several writers note the interrelation between women's later mobilisation in the US war, their politicisation, the eventual success of the war, and land reform campaigns

(Chaliand 1969; White 1982; Tétreault 1994). Women often emerged as radical activists in anti-landlord campaigns, protesting against sexual as well as economic abuse (Eisen 1984). During such campaigns, women were sometimes elected to village leadership posts, an entirely new situation.

The second stage of land reform in North Viet Nam involved the formation of producer cooperatives from 1960 on. The movement for collectivisation was led by the Communist Party, with little democratic participation but was nonetheless relatively non-coercive. Women, especially young women, were among the first to join the cooperatives. White points out that it is not coincidental that the new marriage laws were introduced at the same time that cooperativisation campaigns attempted to widen the agrarian unit (1987: 229). Women who were trapped in arrangements such as marriage to a young boy (for his family's use of her labour as daughter-in-law) joined in order to gain independence from in-laws. In cooperatives, work points were awarded individually so that—as elsewhere—women's labour contributions became more visible and acknowledged.

Women nevertheless remained disadvantaged vis-à-vis men. They were responsible for domestic labour and much work in garden plots, but their efforts were considered less valuable in terms of work points (Eisen 1984; Wiergsma 1988). One day's transplanting, for instance, earned a women ten points and fertilising earned eight points. Men's work earned more: 12 points for a day's ploughing and 14 for a day's carpentry (Tran 1999: 99). In some cooperatives, however, women's tasks earned maximum work points (Chaliand 1969; Houtart and Lemercinier 1984). More commonly, most work done by women was considered light or less skilled and earned fewer points. Meanwhile, men remained more reluctant to join cooperatives due to the threat to their independent status and to their control over women. In one village, the difference between the sexes was so marked that village land was divided into two, with most women joining the cooperative and most men remaining outside (White 1982: 47).

Despite the disadvantages to which women remained subjected, the American war may have been as important as collective land ownership in shifting gender roles. Women performed 65 per cent of agricultural work (Tran 1999: 100) and of necessity, constituted the main labour force during the war. White (1989) critiqued this economic activity as both enforced and as constituting a double burden for women, since the state never took on social reproductive tasks in their entirety. An alternative view is that although women did not achieve equality during the war, social policies did have some success in raising their status (Korinek 2003: 89). It is a common 'trade-off' for women's status to rise only with large increases in workloads. Women moved into diverse economic roles, backed up by the state which absorbed some tasks of social reproduction such as childcare. For instance, a vestige of the war is that women in North Viet Nam are usually the people operating irrigation pumps. Elsewhere, this is a near-universal male activity (Korinek 2003: 94). In the domestic realm, Lilejström *et al.*

(1998) have described how collectives in a forestry area in the far north were not only units for production but were ways to arrange housing and social conditions; they contained as well as the long houses (collective, but poor quality housing) offices, schools, nurseries, and clinics. [2] Thus, some women were able to live partially outside the lineage system.

Cooperatives predominated, but conflicts between the collective and family economies were apparent. The collectives lacked the administrative capacity to supply many basic necessities. Unlike in China, the plan was never for them to supplant individual household production entirely. The importance of the family economy was most visible in the persistence of garden plots. Although some cadres did express concern about renewal of capitalism in the countryside, connections were never made between the petty capitalist and the patriarchal nature of the family farm unit. Wiergsma (1988) gives a powerful analysis of male peasant influence on the reconstitution of patriarchal authority; she uses the term in its classical sense—'rule of the father' (see Chapter 2). She also links the preservation of patriarchy to the preservation of a middle peasantry which remained influential at local party levels and which resisted collectivisation.

After the war ended in 1975, changes occurred within northern cooperatives. Many female managers were replaced by returned male officers who considered it demeaning to be directed by women (White, 1989). 'Higher' level, more technical work came to be more dominated by men into the 1980s. The state's intention had been to further collectivise agriculture; however, agricultural productivity, which had risen, fell. In the south, as noted, collectivisation was much more limited. Falls in agricultural production in the north meant that the southern system in which household services were contracted to the cooperative, continued.

The US war, of course, had devastated the country and led to the deaths of at least one million people (Korinek 2003: 264). It caused ecological and economic devastation, mass displacement, destruction of infrastructure, and lasting ecological damage (including damage to babies and children) due to deliberate use of the toxic Agent Orange, all with long-term effects, not least for women. This war was followed by other, although less pervasive, conflicts in the 1980s with China and then Cambodia. The effective US embargo and withdrawal of Soviet aid inflicted more economic and political damage, so that discussion of Vietnamese economic performance must be understood in this framework. Nevertheless, the mixed popularity of collectives was a factor in lack of growth of productivity in agriculture. Kerkvliet analyses the peasantry's resistance to collectives, although his account does not discuss gender. Male peasants certainly resisted collective production. Although it is likely that fewer women resisted, a number are likely to have joined their male family members. Resistance took the form of everyday actions such as cutting corners in fieldwork and appropriating small amounts of collective land (Kerkvliet 2006). Practices such as foot-dragging and overuse of water buffalos as draft animals (Houtart and Lemercinier 1984; Scott 1985) were also common. These helped to hasten the fall of collectives.

DECOLLECTIVISATION AND *DOI MOI*

A new subcontracting system was instituted in agriculture in the north from 1981 (Tran 1999). Similarly to the system existing in the south, the cooperative contracted for the delivery of final products with individual households or groups of households. The change was precipitated by several factors, of which similar changes in China from the late 1970s (see Chapter 5) were most important. The need for economic recovery after the war also played a part, as did peasant resistance. In this regard, men's dislike of loss of control over wives' and daughters' work and persons was a significant factor (Wiergsma 1988). As elsewhere, the return to a family labour system has meant that women's work became less visible, though the advent of the subcontracting system meant that women could again assume traditional roles as market women (White 1989: 187).

By the late 1980s, decades of war had meant a general deterioration of conditions for rural women. Customarily, an adult male and female, a water buffalo, and a plough were the ingredients of a farming household. The reality, however, was often that the man might be dead or absent; the buffalo, dead or sold, and the plough, too expensive. A large minority of women were, and are, left to farm on their own, lacking even draught animals (White 1989).

From 1988, household rights over land were greatly strengthened and a much fuller decollectivisation was initiated in the doi moi policies. Cooperative lands were leased to farming households for ten- to fifteen-year periods, although in practice this period was often shorter, sometimes three years (Tran 1999: 101). State proposals initially favoured 'strong' or wealthier farmers who were able to contract more fertile land for larger production quotas. Because of widespread opposition in peasant villages, this proposal was dropped and by 1994, principles of equitable distribution were reinstituted (Luong and Unger 1998: 65). As in China's decollectivisation, doi moi tended to generate very small and scattered plots of land, meaning that cultivators had to spend time travelling between them (Thi cited in Korinek 2003), and some cadres appear to have retained privileges in land allocation Korinek (2003: 145). Luong and Unger (1998: 86) stress that redistribution was equitable in gender and age terms, and this was indeed the stated policy. The reality, however, differed in a number of villages (Tran 1999). Women received less land than men for several reasons. For instance, women retire at 55 years rather than age 60 and thereafter receive half shares; men commonly hold state-contracted land, but women rarely do (Tran 1999: 103).

In 1993 a new land law was enacted, giving longer periods of use rights of up to twenty years for annual crops and to fifty years for perennials such as trees and coffee bushes (Tran 1999: 101). The amount of land allocated per individual differs by region, with the lower per capita allocation in the Red River Delta and the highest levels in less-populated regions such as the Central Highlands and northern mountain areas (Kabeer and Tran 2002). In effect, a land market was permitted (Dao 1995: 157) as use-rights can be transferred between individuals and families (Gammeltoft 1999: 30). Landlessness and differentiation became more of a possibility. By the late 1980s, much land in the south had been returned to pre-land reform or collective owners, and a sizeable class of rural landless had emerged (Kabeer and Tran 2002; Korinek 2003: 62). In the north, landlessness remains much rarer. However, the poorest risk confiscation of their land if they cannot reach production targets (Gammeltoft 1999).

Land rights certificates are issued by local authorities who should allocate land to women as well as men. Women whose names are listed on certificates should have their land rights protected in cases of divorce or widowhood. However, men are often the only people named on household certificates (Gammeltoft 1999), meaning that wives' rights to land and housing may become highly contingent. In four northern and southern villages studied by Kabeer and Tran (2002: 183), land had been registered in most households only in the man's name. This practice follows an international trend. Kabeer and Tran identified co-ownership by husband and wife in the two southern villages studied as 1 per cent and 3 per cent. Co-ownership was far more common in the north, at 20 per cent and 27 per cent, respectively, in the two villages studied. Many of the women respondents in their research complained that the distribution process had been unfair, and that the land committees—mainly made up of men—had not consulted them. A number of women also said that they had been allocated much smaller holdings than men. As in China, the extended land leases and curtailing of land reallocations has also led to problems for women who move outside their natal villages upon marriage, as they may have to forfeit their use-rights (Kabeer and Tran 2002).

Land legislation enacted in 2003 strengthened women's rights and *requires* that husbands and wives' names be listed on the land use certificates (ActionAid 2005: 70). The new law, however, has not been implemented. Divorced women remain likely to lose their land. In cases of widowhood, the son's name is more likely to be entered on use certificates than is that of the widow (ActionAid 2005: 71).

In Viet Nam, decollectivisation was very rapid and a near-Chayanovian peasantry (see Chapter 2) has been established (Watts 1998). Land ceilings, at least officially, are very low at two to four hectares. Land has been more thoroughly returned to household control. Interestingly, the size of land parcels corresponds roughly to those in the nineteenth century, particularly in the north. The state seems to have as a model the successful agrarian reforms of South Korea and Taiwan in the 1950s.

With doi moi, other features of the collective period were curtailed or ended: the cooperative structures were in the main liquidated, and their support services and state marketing structures abolished (Watts 1998). The household registration system which had controlled migration was loosened and so migration became very common. Korinek comments that nearly every household in her Red River Delta study had a labour migrant member (2003: 112). The marketing sector and informal economy, as noted, have flourished. In general, productivity rates and food consumption have risen with doi moi (Kerkvliet and Porter 1995; ActionAid 2005), and child malnutrition is much less common (Korinek 2003). Viet Nam has become the world's second largest exporter of rice, and exports of other crops such as tea, tobacco, and coffee have risen dramatically (Luong and Unger 1998; Korinek 2003: 145). Southern and central Highlands regions have been particularly productive (Luong and Unger 1998). Annual growth rates in gross domestic product rose from 2.3 per cent in 1980 to 8 per cent in 1989; the extent of severe poverty decreased from 70 per cent in the mid-1980s to 55 per cent in 1993 (Kabeer and Tran 2002: 109). Improvements in productivity are due to intensification by multicropping and use of increased inputs and to the greater care with which plots are tended (Kabeer and Tran 2002).

Viet Nam remains a very poor country, nevertheless, with *per capita* incomes of $690 in 2006 (World Bank 2007), with rural incomes considerably lower. Despite state efforts, industrialisation has not occurred on the Chinese scale (Luong and Unger 1998). Severe and chronic poverty is particularly a rural problem (Kabeer and Tran 2002); the poorest villages tend to be remote rural ones solely dependent on agriculture (Korinek 2003: 147). In this context, it is difficult to base a livelihood solely on agriculture, so households try to diversify to minimize economic and social risk in a context of great impoverishment (Gammeltoft 1999; Kabeer and Tran 2002; Scott 2003).[3] Diversification involves both men and women, although in different ways. Household survival often depends upon women's activities, whether as farmers, operators of small enterprises, wage labourers, or 'helpers' in small businesses. Women's responsibility for household survival weighs heavily upon them, often affecting their well-being (Gammeltoft 1999: 33).

Women and *Doi Moi*: Losses and Gains

Gendered shifts in economic activity and family relations have taken place with doi moi. An underlying question is whether any losses for women are attributable to liberalisation or alternatively to a return to tradition. Decollectivisation, as stressed, means that women spend more time on household-based economic and reproductive activities. Given Viet Nam's peasant base, this for many means farming and trading. Some women do work for wages, but this is mainly a correlate of poverty (Kabeer and Tran 2002: 139). One of the most important effects of doi moi has been the opening up of opportunities for trading and operating small or micro-enterprises, allowing women to return to such activities (White 1989), with which they are more commonly engaged than men (Korinek 2003: 404).[4] Such enterprises, given diversification, are usually managed alongside agricultural activity.

The main livelihood basis, however, continues to be in agriculture, especially since liberalisation and foreign direct investment has not resulted in gains in formal employment (Jenkins 2006). Some sources mention, as in China, eastern Europe, and elsewhere, a feminisation of agriculture (e.g. Bich 1999: 97; Tran and Le 2000: 100) and of related sectors such as forestry work (Lilejström et al. 1998). Certainly, many men migrate, sometimes to different provinces, and this has left rural women more responsible for agricultural activities. Korinek's careful study of the Red River Delta, based partly on the Vietnamese Labour Survey (VLS) data, indicates another scenario (2003). She found an 'agriculturalisation of the labour force' rather than feminisation of agriculture, chiefly a consequence of military demobilisation of men. Of those young people who remained in the area rather than migrating, work had become more concentrated *within* agriculture.

Womens' workloads after liberalisation also depend on factors such as childcare. Viet Nam, like China, has a state family planning or population limitation programme. The programme allows one or two children per household and three for ethnic minorities; most rural families have two children and these should be spaced three to five years apart (Gammeltoft 1999). The programme is much less rigidly and coercively applied than in China and relies on campaigns such as 'happy families' to convince people that household wealth and well-being will be enhanced with small families. The number of children per woman has declined from 6.1 in 1960 (Gammeltoft 1999: 18) to 2.3 in 2002 (World Guide 2005: 606). The decline in numbers of children eases work in childcare and housework.

In many respects with doi moi, material life for rural women has become easier; more food is available and possibilities of acquiring some consumer goods are enhanced for many groups. However, a countervailing trend also integral to neoliberalism is reduction in subsidies for healthcare and the introduction of fees for education and healthcare. The ability to cover such fees is of major concern to many, including most rural women.

In Viet Nam as elsewhere, women are more likely to plough resources back into the household (Kabeer and Tran 2002). Healthcare fees impact heavily on incomes of the rural poor (Bélanger 2002: 330), and gaps in access to health services have widened in recent years (World Health Organization [WHO] 2006). Gender differentials in family educational expenditure have also re-emerged in the doi moi era (Korinek 2003: 160). Bich writes that both sexes of children are sometimes withdrawn from schooling due to increased needs for their labour; though this is more marked for girls (1999: 90; Bélanger and Liu 2004). Such practices may have consequences for the country's literacy rates, which stood at 86 per cent for women in 2000 (United Nations Educational, Scientific, and Cultural Organisation [UNESCO] cited in ActionAid 2005: 67). The gap between men and women's literacy for those aged over 25 years was 15 per cent (ActionAid 2005).

Women face increased pressures on workloads under doi moi. Nearly all residents of the Red River Delta, for instance, work very long hours (Korinek 2003), and this is compounded for women, given their duties at home. Male migration in many cases also means that women must work harder in agriculture (Bich 1999: 97) as well as in small enterprises. Other observers note that the policy changes of decollectivisation have brought both gains and losses for women, as accompanying welfare systems have been dismantled. One woman interviewed by Gammeltoft said 'Before we had more time to rest and to talk to other people—it was more happy [*vui* 'joyful'] to go to work because everyone worked together.' (Gammeltoft 1999: 32). Today, people's work is arranged more autonomously but work is more demanding and time consuming. Wives' retreat into the household means that there is less contact with others. Because women realise that they may be responsible for family welfare, they are often impelled to work to their limits.

The market reform period has also meant a contraction in women's representation in higher status positions (ie teachers, administrative workers, and healthcare workers), and younger, married men have moved into this work (Rama 2002; Korinek 2003: 410). Today success or failure is seen as a personal responsibility, whereas previously everyone [in the north] was equally poor. New opportunities exist now but so does greater risk and uncertainty, accompanied by increased stress (Gammeltoft 1999: 34).

In summary, then, with doi moi there has been a move for women to be confined, or more confined, to household-based activities, raising the question of: *how* the shape of and relations within the household may have been altered? In the sphere of family law, regulations benefit women. In particular, the 1986 Family Law, drafted by the WU, affirms joint control of household property, joint consent to economic transactions, and equal household domestic responsibility. It also attempts to protect women against divorce during pregnancy and from spousal violence. In theory, a husband (or wife) convicted of violent abuse may be imprisoned (Tétreault 1994). Although it is highly unlikely that such provisions operate fully in the countryside, they have raised rural as well as urban women's legal status.

Households today take both extended and nuclear forms; though nuclear families now predominate, constituting 71 per cent of households (FAO 2002). Upon marriage, most couples live with the husbands' parents but then set up their own household. Household division into nuclear families has increased due to land policies; couples remaining with parents will not receive farmland or house plots (Tran 1999: 110). Bich writes that the conjugal relation still remains subordinate to filiation. The purpose of marriage is to produce offspring, and elders are to be venerated (1999). In 1987, mother-in-law control over daughters-in-law had only slightly loosened (Wiergsma 1991). Official policy, building on the idea of the 'three obediences', stresses women's but not men's special family obligation as well as their responsibility for building socialism. Gammeltoft comments on women's responsibility for ensuring family harmony; as in China, this is deeply internalised and usually enjoins compliance or at least adept management of conflicts (1999). It is possible, however, that increased importance of nuclear families will increase wives' influence (Jacobs 1995).

Countervailing trends are evident. On one hand, with decollectivisation there has been an increase in attention (and expense) paid to weddings and funerals (Sikor 2001: 945). Collectivisation often confiscated ancestral mounds, and religious rituals including ancestor veneration have been revived (Bich 1999). A 2006 report indicates that the sex ratio is altering to the detriment of girl babies (Sabharwal and Thien 2006). As elsewhere, the introduction of ultrasound technology has facilitated sex selection of boys. Because this is available at present mainly in urban areas, the declining sex ratio at birth is not a rural phenomenon, but it may become so. Again, this trend indicates both the importance given to sons in carrying out ancestral rites and concern about economic security in old age.

The incidence of female-headed households has increased in Viet Nam and although an international trend (Chant and Campling 1997), this is distinctly 'non-traditional'. In 2005, 17 per cent of rural households were classed as female headed (FAO 2002; see also Scott 2003).[5] To some extent, this is a legacy of male mortality during the long period of conflict and war. Many female-headed households result from male migration and the advent of 'visiting' marriages. Lilejström et al. note that family separation may at times be another strategy to minimise risk: one spouse (usually the woman) might retain access to land in the home village while the other (often the man) might work away as a state employee (1998:48). A new phenomenon is that single women have established their rights to be mothers and to head families (Liljeström et al. 1998). The established practice is that the child takes the mother's name and the identity of the father is never revealed (Bich 1999: 87). This practice can cause hardships but is seen by many, including the WU, as preferable to the return of *de facto* concubinage (Lilejström et al. 1998). Single mothers have rights to house and land but in practice, are among the worst off of peasant households. They have rights to only one adult land allocation and may receive inferior land. Disadvantaged access to land has been identified as a major factor explaining the poverty of female-maintained households (Kabeer and Tran 2002: 135).

It is an advance that unmarried women, widows, and women with absent husbands, about 20 per cent of rural households, are able to receive land use certificates in their own right (Tran 1999: 109). However, rural practice continues to disadvantage women in their access to land. Divorced women encounter major problems, particularly if they originate from 'outside' villages losing their use-rights unless they remain in the ex-husband's village (Tran 1999: 112). While a divorced woman should receive compensation for any loss of land rights, it is difficult to estimate fair compensation and women are often underpaid, as well as being left landless (Tran 1999: 112; Scott 2003). A married-out woman may also be disadvantaged if, for example, she has to walk or cycle to her own village daily to work land in order to keep her rights 'active' as well as to tend crops (Gammeltoft 1999: 32). With the partial retreat of the state, there is more reliance on kin networks (Bélanger 2002), and land inheritance is kinship mediated. This may mean that sisters have to assert rights against brothers, risking alienating them and countering women's 'responsibility' to maintain harmony. Women are far less likely to inherit agricultural land than men (ActionAid 2005), and they rarely inherit forest land (Scott 2003). Thus, many gaps in the law exist, and it appears that the complexities of individual use rights coupled with gender rights were not fully considered in drafting laws. In the process, many groups of women may lose out.

CONCLUSION

Viet Nam presents a highly unusual case of an egalitarian decollectivisation process. Conditions in the north approximate a Chayanovian peasantry, with relatively little differentiation. Rural women in Viet Nam have gained some benefits from decollectivisation: they have been able to take up traditional marketing roles and to add to their own and to family income. They have also gained more flexibility over their own work schedules and pace, although not over the overall amount of work. Increased prosperity and food security is likely to have benefited women as well as men. Here, the egalitarian nature of decollectivisation means that households have benefited more evenly than in most cases.

For many, however, losses outweigh gains. Decollectivisation, as in other examples, has meant loss of visibility and of individual remuneration for women's agricultural work. Wives have become more confined to households, and husbands and fathers have gained influence—although women's trading activities to some extent counter this. Increased workloads for women, isolation, and increased responsibility for family welfare are reported, since collective and state services have been withdrawn. Lineage practices, too, are becoming more common. Coupled with Confucian ideas about women's roles, these undermine women's household influence and autonomy. In Viet Nam, however, the increase in nuclear families and of female-headed households are countervailing influences. So too, is the continued influence of the WU and the strong legal base for women's rights in the marriage law.

Vietnamese women now have strong legal rights to land and to property. However, the law has not translated into practice and women remain disadvantaged in terms of landholding in the new, market-based situation. As in China, the lack of civil society women's organisations and of other social movements is a great hindrance. It is important not to jettison the support and achievements of state-backed feminist organisations. The WU has been extremely proactive in lobbying for women's rights. Particularly in market conditions, however, quasi-governmental organisations are not able to substitute for autonomous women's movements. At the same time, it is well to recognise that feminist movements and networks are nearly always urban rather than rural (Jacobs 2004b). Rural women in Viet Nam do not lack agency, but they do lack autonomous movements that might complement the WU. Women's rights in emotive areas such as land are rarely gained without outside support from state bodies and from social movements.

Part III

Household Models of Reform and Alternatives

7 Mobilisation and Marginalisation
Latin American Examples

This chapter discusses three Latin American examples of agrarian and land reform: Mexico, Nicaragua, and Brazil. These comprise important cases from different parts of the region and with different trajectories, but all indicating the relatively limited nature of Latin American land reforms. This situation reflects either lack of redistribution (Brazil), lack of support and constraints following land reform (Mexico), or the advent of war and counter-reforms soon after (Nicaragua; see also Chapter 6). These examples also indicate the general exclusion and marginalisation of women within the region's agrarian reforms (Deere and León 1987, 2001), signalling both weaknesses in policy and the strength of domestic ideologies and *'marianista-machista'*[1] gender regimes. The experiences also differ from those discussed in the previous section. In most of Latin America, land reforms were organised predominantly along individual household lines or else with a mix of collective and individual household tenures, but always co-existing with a large capitalist sector.

The examples firstly discuss the general background and trajectories of agrarian reform and then gendered effects, as well as effects of counter-reforms and land titling.

MEXICO

Mexico constitutes the earliest example of agrarian reform in Latin America and one of the earliest far-reaching reforms in the world. In 1910, when the uprising began against the long-lasting authoritarian regime of Porfirio Díaz, 1 per cent of *hacendados* (landowners) controlled 97 per cent of land, including 40 per cent owned by US citizens (Thiesenhusen 1995: 35). Meanwhile, 92 per cent of the rural population, 15 million people, were landless (King cited in Sobhan 1993: 36; Thiesenhusen 1995: 30). Peons (tied peasant cultivators) had rights to cultivate small plots on the edges of agricultural estates but were bound in debt servitude. Usually this lasted throughout their lives, through accumulation of debt through credit. Normally, peons worked six days for the master and one on the family plot.

Women usually performed household tasks for landowners, and sometimes were forced to offer sexual services. The hacendado exercised police, and often, magisterial powers so that the hacienda bore some similarity to feudal estates. The 1910 revolution drew in peasants and rural workers under the leadership of Emiliano Zapata and Pancho Villa, who began to seize haciendas in a violent insurgency (Wolf 1969). The *Plan de Ayala* which promised land reform was agreed in 1915 by Venustiano Carranza, head of the pre-Constitutional government and later, president. However, by 1923, 59 per cent of households with an average of three hectares still owned only 0.8 per cent of all land (Sobhan 1993: 36). US citizens still held 20 per cent of land (Thiesenhusen 1995: 35).

The revolution was, and is, the bloodiest conflict to date in the Western hemisphere. By 1920, some 1.5 million of the 14.5 million population lost their lives (Hellman 1988: 3). One outcome was the Constitution of 1917 whose Article 27 declared land a resource of the state which had the right to assign it to individuals or as social property. Although some land was redistributed throughout the 1920s, the major agrarian reform was instituted under the socialist-leaning president, Lázaro Cárdenas (1934–1940). Some redistribution continued until the early 1970s as a result of continued peasant agitation (Dawson 2006). By 1970, 43 per cent of all farmland was redistributed to 66 per cent of rural families (Sobhan 1993: 36). In the period immediately following Cárdenas, efforts were made to improve rural infrastructure, especially through irrigation and road building.

The Mexican agrarian reform was one of the largest to date outside socialist countries, but it still did not cover the whole agricultural sector. In 1970, 10,000 agricultural holdings of over 1,000 hectares still owned 32 per cent of all farmland, indicating the size of ranching haciendas and the predominance of commercial agriculture, especially in the north (Sobhan 1993: 37). The land ceiling imposed on larger farms was generous, being between 100 and 200 hectares of irrigated land (de Janvry et al. 1997: 284).

Land was redistributed in nearly 30,000 communal units (Dawson 2006: 35) called *ejidos*. These are owned by the community although usually worked by individual households. The community would assign usufruct rights to households which had to be passed on to family heirs. They could not be subdivided, sold, or rented outside the ejido—not legally. However, an *ejidatario* not cultivating his or her land for two consecutive years would lose it. Up to twenty hectares of land could be cultivated, but inequalities in holdings did and do occur within ejidos.

In principle, the ejido is democratically organised on the basis of one person: one vote with a three-member rotating governing board, a vigilance committee, and a monthly general assembly. In practice, a pattern of patronage and *caciquismo* (influence mongering) deeply rooted in Mexican history is often replicated within ejidos (Brunt 1992: 41). The presidents of ejidos tend(ed) to monopolise decision-making and to act as patrons and brokers. The state and the ruling party, the *Partido Revolucionario Institucional* (PRI) controlled the flow of resources to ejidos, so the presidents became clients of bureaucratic superiors. These might deliver to the ejido piped water or land grants. In turn, ejidatarios were dependent upon the president and other officers for patronage (de Walt et al. 1994). Thus, the corporatist organisation of the PRI and of the Mexican state meant that peasants had to resort to party and state agents as intermediaries between themselves and national society. A strong argument exists for seeing ejidos at least in part as mechanisms of party control (de Janvry et al. 2001: 286–87).

Debate exists concerning the productivity of ejidos. State inefficiency and corruption often stifled agricultural innovation. Nevertheless, by the 1960s, Mexico had achieved food self-sufficiency (Green 1996: 267). At this time, however, spurred on by international agencies, production of export crops was favoured and Mexico became a net importer of its basic foodstuff, corn (maize). Although much credit is given to the success of commercial farms, the ejidos' contribution to economic development was not always acknowledged (Thiesenhusen 1996). A major problem for ejidatarios was the inability to obtain credit without land titles (Myhre 1996). This problem was only partly solved by the undercapitalised rural credit bank, BANRURAL (Thiesenhusen 1995: 42). Private farms were favoured in obtaining credit and were more able to employ green revolution technologies. The two agrarian systems coexisted in a bimodal structure, in which ejido agriculture had little benefit from investment, credit, or new technologies. Thus, the 'playing field' was highly uneven (De Walt et al. 1994).

In general, agricultural production declined. This was the case from the 1970s and according to some authors (Arizpe and Botey 1987: 73), from as early as the 1950s. A combination of rising population, stagnant agrarian reform, and new opportunities created by industry as well as migration to the USA, meant that by the 1980s, livelihoods, as elsewhere (Ellis 2000; Jacobs 2002) became diversified with many households relying on a combination of activities for survival. Long-distance as well as more local migrations became common strategies, with (at first) unmarried daughters travelling to find work. By the 1960s, sons and husbands also began to migrate, often leaving middle-aged or older women behind to farm (Arizpe and Botey 1987: 79). Women, too, entered rural labour markets, often remaining at home to work, for example on piecework. However, some became involved in seasonal migration and subsequently, a number worked in agribusiness or in border export processing zones (EPZs). Nevertheless, title to the ejido plot remained important for subsistence and sometimes symbolically (Brunt 1992; van der Haar 2000).

In 1992, the government ended the agrarian reform process definitively, repealing Article 27 of the 1917 Constitution. An evaluation of the agrarian reform process would concede major weaknesses, such that many beneficiaries became disillusioned and the sector, bedevilled by government inefficiency, fell short of its producer potential. However, there were many strengths. The reform served as an incentive for small producers to become a major source of staples such as beans and maize. It also reduced inequality in the distribution of resources, particularly before the 1970s (Thiesenhusen 1996: 44). As a result, some rural producers who otherwise would have migrated to cities or to the USA remained on the land. Where migrants were unsuccessful in the USA they could be reabsorbed into the ejidos. Lastly, whether or not seen as a 'strength', land reform helped to quell rural unrest.

Gender and the Mexican Agrarian Reform

Women were marginalised within the Mexican agrarian reform from the beginning. The 1920 ejido law states that land should be distributed to *jefes de familia*, normally taken to mean the male household head (Stephen 1996: 291). A 1927 law interpreting the implementation of Article 27 (Article 97 of the *Ley de Dotaciones y Restituticiones de Tierra*) specifies that those eligible to be members of an ejido should be 'Mexican, males over 18 years of age or single women or widows supporting a family.' Thus, men qualified on the basis of gender alone whereas women qualified only by virtue of supporting family members in the absence of a man, as widows and temporarily, in lieu of sons (van der Haar 2000). From the 1930s, feminists began to agitate for land rights for rural women (Deere and León 2001: 70). A 1940 code allowed *ejidatarias* (female members of ejidos) to rent or sharecrop land if they had small children. A 1943 code, in effect until 1971, provided that if a man did not work the land for two years, the parcel would revert to his family rather than to the ejido commission (Deere and León 2001: 70). However, it was not until 1971 under Article 200 of the agrarian reform law that women qualified for ejidatario status on the same basis as men (Stephen 1996: 292). By this late stage, few could support themselves by agriculture alone and most people had developed diversified livelihoods. By 1984, female ejido members accounted for 15 per cent of the total (Arizpe and Botey 1987) and 17.6 per cent by 2000 (Deere and León 2001: 73). Most of these were elderly widows, since the main method of land acquisition was inheritance. Many widows were unable to cultivate their own land and it was left in the hands of a son or brother (Arizpe and Botey 1987: 71).

The 1971 law had an unusual feature: it mandated the creation of agro-industrial units for women (*Unidad agrícola industrial de la mujer*, or UAIMs). A UAIM only had land equivalent to that of one ejidatario, regardless of the number of women. Although 8,000 UAIMs were set up legally, by 1987 only 1,224 were operational and most were not commercially viable (Arizpe and Botey 1987: 72). Some UAIMs raised crops; others were engaged in poultry raising, sewing, or embroidery. In some cases, especially where land was scarce, men refused women rights to the collective plot. Nevertheless, UAIMs were useful in raising awareness of the need for women's employment.

Several accounts indicate the symbolic and procedural means by which women have continued to be marginalised within ejidos, either as ejidatarias or as wives or daughters of members. A summary of anthropological studies notes that women take little part in communal assemblies, and that their rights and legal authority over land have not been implemented (De Walt et al. 1994: 37). Despite their marginalisation, women displayed strong interest in acquiring land and in land issues (Brunt 1992). By 1998, 20 per cent of ejidos had elected a woman onto the board (Deere and León 2001: 73). These general points can be explained more fully by examining case studies of gender relations and women's status within ejidos.

In El Tule ejido near Oaxaca, about half of female ejidatarias were widowed or single mothers. The other half received or inherited rights because they had no brothers or else their brothers had migrated permanently (Stephen 1996: 297). Where women held land, this was generally in small plots of under two hectares so that they had to engage in additional income-generating activities. Many with land were unable to cultivate it because they lacked skills such as ploughing and were unable to hire in labour.

An ethnographic study of a well-off ejido family indicates the power the husband wielded over the wife (Nuitjen 1998). The wife of the ejidatario had raised many children, was highly religious, and therefore conformed to the marianist image of motherhood. Despite this, she was not allowed to leave the home without the husband's permission, a semi-confinement to the domestic arena that occurs in many village settings. In this account, the mother did wield authority when she could, but her influence was confined mainly to her own sons and in respect of their treatment of *their* wives. Even in the late twentieth century, women who had migrated to the USA and who had different ways of behaving, nevertheless upon return to the village had to conform to village mores, including matters such as modest dress, domestic semi-confinement, subservience to husbands, and toleration of the double sexual standard. Women flouting village norms might risk gossip or a bad reputation.

Brunt's (1992) study of 'El Rancho' in Jalisco terms ejido women a 'muted group' due to their silencing as well as the structural constraints they face. As is common elsewhere, wives often found that husbands spent household income on drink and other women rather than reinvesting it (Brunt 1992: 128). Thus, women said that their own lives improved with the advent of commercial production of melons and cotton, since they had more control when they received a wage. But this also caused conjugal conflict: a number of husbands refused to allow their wives to work and so the wives could only do so when husbands were away (Brunt 1992: 130–31).

Overall, however, agricultural modernisation has increased the tendency for men to be seen as 'agriculturalists' and women as 'housewives'. Both men and women in the 'El Rancho' study express the domestic as an ideal for women's behaviour. Women should not claim land, and agriculture is 'men's business'. To claim their rights, women have to project themselves as responsible mothers thinking of the future of their children, and particularly, their sons. This is the only acceptable rationale for women's claims, and, in practice, even widows or deserted women able to mount such claims must depend upon their ability to manipulate patronage relations. Benefits of holding land exist, however, even in the constrained circumstances many women face. Having legal title to land offers financial control and the status to operate in a man's world (Brunt 1992: 184). Gender disadvantages remain, but there is increasing room for manoeuvre.

A Nayarit study describes difficulties women had during the 1980s in trying to organise autonomously. This attempt took place within the Women's Council in an ejido union, Lázaro Cárdenas Ejido Union (Stephen 1997). The Women's Council depended upon the (mainly) male organisational structure, and they themselves also tended to reproduce a similar hierarchical structure in which authority figures' views were followed. Little overt challenge to domestic gender ideology existed, defining the women as wives and mothers whose roles were to improve family welfare. Nevertheless, participation in the Women's Council and in income-generating projects did change women's consciousness and gender relations. Women struggled for simple representation within the ejido union and for rights to leave the domestic sphere to carry out their projects. Not infrequently, husbands opposed them with violence (Stephen 1997: 190). The encouragement to form the Women's Council came from outside, through urban women organisers. This perhaps indicates the difficulty of organising on the basis of gender within village settings.

A study of a Morelos ejido (Martin 1994) focuses on political struggles. In a complex process, women initially came together in an extra-ejido women's organisation, the Buena Vistan Women. They organised outside the ejido due to their difficulty in finding voice within ejido assemblies. Their organisation positioned itself as standing for community against the class politics voiced in ejido assemblies. However, and perhaps ironically, the outcome simply reinforced class splits. The community organisation eventually dwindled to a few women entrepreneurs and businesswomen, meanwhile, poorer peasant women continued to see their interests as lying in the ejido assembly and as struggling for better representation within this.

These examples highlight several processes affecting women within ejidos. Firstly, as stressed previously, the overall context is one of marginalisation. This is true especially for younger women, given that women sometimes gain in status over their lifetimes. Most women who become ejidatarias do so as widows. The strength of domestic and marianist ideas and images continue to be strong in rural areas, constraining women's participation (particularly, in spatial terms) as well as their assertion of rights. Women often face difficulties being taken seriously within ejido assemblies, although at least some manage to overcome such attitudes and even to be elected onto governing councils. In terms of wider political struggles, women's politics sometimes took community forms, reflecting in part the difficulty of organising within ejidos. Lastly, when women did succeed in becoming ejidatarias, they became relatively privileged, although usually less so than male ejidatarios.

The Mexican Counter-reform and Privatisation

Agitation to disband ejidos and to privatise began in the 1980s. It developed in the wake of the general neglect of ejidos since the 1970s as well as Mexico's debt crisis in the 1980s (van der Haar 2000). The land reform was ended in 1992 with alterations to Article 27 of the Constitution. Among other changes, the decree abolished most restrictions on maximum size of holdings, allowed private companies to buy agricultural land, and allowed ejido farmers to sell plots. These measures accorded with the overall liberalisation and deregulation of the economy. The North American Free Trade Agreement (NAFTA) of 1994 was a further push to 'modernise' and to open up the economy for exports. Thus, one purpose of ending land redistribution was to create a land market and to open up the agricultural economy for national and international capital. Export cropping was to be further encouraged, and institutional and financial support for small farmers was further curtailed (Green 1996: 267–68).

The change in the Constitution terminated the role of the state in land distribution. Privatisation of the ejidos was permitted, except in Indian [native American] communities (Brown 2004: 20–21). Ejidos which opted for change could vote to disband or to distribute individual titles either to all or to some members. Ejidatarios can now enter into joint ventures with outside bodies, including outside capital. The Program for the Certification of Ejido Land Rights and Titling of Urban Household Plots (PROCEDE) was created to deal with titling; other bodies were created to mediate conflict and to deal with land claims.

In some Latin American contexts, women have made gains within neoliberal land titling legislation, which have sometimes mandated joint titling for husbands and wives (Deere and León 2001, 2003). Mexico, however, constitutes an exception (Hamilton 2002: 120) since its counter-reform failed to provide for joint titling or for the needs of single female household heads. Women's rights have been eroded by the neoliberal reforms in various ways. Firstly, all major decisions have to be taken by the ejido assembly, from which women are largely excluded and therefore unable to participate in deliberations over the legal framework. Secondly, land changes from being a family resource in the ejido to one that can be disposed of at will by the ejidatario (Deere and León 2001: 250). Although the ejidatorio's spouse and children have first buyers' rights, they have only thirty days to make arrangements to purchase the land, an impossibility for most women (Stephen 1997). Thirdly, the creation of UAIMs is no longer mandated within the ejido (Deere and León 2001) because the redistribution of land has been halted. Fourthly, ejidatarios can now choose their heirs: formerly, choice of beneficiary was confined to family members. Thus, land can now be passed on without reference to the wife (or indeed, the husband) or children.[2]

Privatisation as entailed in the revision of Article 27 tends to disadvantage the poorest cultivators, among whom women are disproportionately represented. In Mexico, this occurs within a context of semi-proletarianisation for most rural women and ejido men. The first years of liberal counter-reform have accelerated social differentiation. For instance, credit is only available to farmers who are already solvent (Appendini 1996: 67). Although a minority of households are successful smallholders, a retreat into subsistence farming has been more common (Appendini 1996). This strategy underlines the importance of land, particularly for food security, but is accompanied by intensification of family labour, wage labour, and migration (Appendini 1996: 66).

More optimistically, however, other observers argue that the weakening of state controls over the peasantry has allowed the emergence of a peasant economy (de Janvry et al. 1997), characterised by reliance on family labour, intercropping, international migration, an increase in cattle raising, and forms of mutual support for access to labour and insurance (1997: 203).

The case of Chiapas and the *Ejército Zapatista de Liberación Nacional* (EZLN) movement is important, but it is not possible here to explore it at length. The initial agrarian reform did include Chiapas state, which has large indigenous populations. However, practices such as debt peonage, forced labour, and landlord and police corruption and violence were more widespread than elsewhere (Burbach 1994; Harvey 1996; Warman 2003). Landlordism, police brutality and official indifference, as well as continued poverty, helped to fuel the population's resentment. Other issues involved in the 1990s uprising included the advent of the NAFTA and Indian identity. The rescinding of Article 27 also played a part, since peasants' maize and cereal cultivation is endangered through the withdrawal of subsidies, credit and technical assistance. Thus, the rebellion was fuelled in part by the exclusionary impact of Mexico's agricultural modernisation and the formal end of agrarian reform (Kay 2001: 758; Harvey, 1996; Paulson 2000; Warman 2003). Indians found a space for autonomous development within ejidos, which was threatened by the repeal of agrarian reform as well as loss of land and community-level control (van der Haar 2000). Large-scale land invasions covering over 500,000 hectares took place in the wake of the Zapatista uprisings. These forced the government to buy land for redistribution *despite* the official end of land reform. In the east of the state, where revolt was most intense, privatisation also had to be postponed (van der Haar 2000: 158).

In general, liberalisation has not signalled the complete demise of ejidos echoing experiences of decollectivisation that will be discussed in Part II. In the mid-1990s, 72 per cent of ejido land was being parcellised (de Janvry et al. 1997: 283). Ejidos emerged from reforms as institutions with new functions such as supporting peasant production. For instance, access to machinery through collective ownership rose overall in 1990 to 1994 (de Janvry et al. 1997: 208). De Janvry et al. argue that a peasant economy inserted into markets may be able to use the comparative advantages that ejidos offer. Other studies indicate that PROCEDE was deeply distrusted and a view that its involvement would be a step towards total loss of land, especially in case of debt (Plaza 2000: 171). A Oaxaca report found that *none* of the ejidos visited were fully privatised (Brown 2004: 21). Privatisation was limited to peri-urban land which had become more valuable. Thus, privatisation was only seen as useful in order to sell land rather than to farm more efficiently. Reasons for not privatising include: the low value of land due to poor soil quality; to prevent land concentration as in the days of the *hacendados*; and the need to preserve a sense of community (Brown 2004: 25). Brown's study concludes that the 1992 reforms had some success in resolving boundary disputes and increasing tenure security. However, women's rights had been undermined.

It is the consensus that women's rights within ejidos—already tenuous—have been further eroded by lack of attention to gender issues within the 1992 counter-reform. Nevertheless, at least two studies have found that some women, particularly widows, have *strengthened* their positions within ejidos, post-reforms. Plaza's research found that women's greater influence is due to feminisation of the agricultural wage-labour force, giving rise to the perception that women are able farmers (Plaza 2000); such a view is relatively new in Mexico. Hamilton (2002) studied four ejidos in the north and centre of the country. Rather than finding an erosion of women's status she found improvement. Women's agricultural and other labour inputs into the farm increased in a situation of economic crisis since the early 1980s. Along with women's social activism, this made their participation more visible. Another important factor was that agricultural production was increasingly subsidised by remittances sent by children to their parents, in a context in which 10.2 million Mexican-born people (of a population of approximately 106 million) live in the USA (Tuirán et al. 2005: 7). Here, mothers have increasingly been seen as 'safer hands' in terms of use of household money and daughters are particularly likely to send remittances to mothers.

Beyond women's enhanced roles in agriculture and in rural households, a more general change in ideas about women's capacities appears to have emerged which may have hard-to-pin-down but important effects. Such perceptions may be influenced by the impact of transnational migration, as both women and men migrate and may 'bring back' new ideas about gender relations (Nuitjen 1998). Lastly, feminist organising within Mexico itself no doubt has an impact (Deere and León 2001). However, none of these positive influences have influenced the crucial arena of legislation, which remains highly detrimental.

NICARAGUA

The story of agrarian reform in Nicaragua is one of countervailing tendencies. Women made gains under the Sandinista revolution; cooperatives were formed; and women's roles in production became more fully recognised. Women were allowed to become cooperative members but often faced opposition from husbands and family members. With defeat of the Sandinistas and the advent of neo-liberal land titling, most cooperatives have been disbanded and women, like men, have lost out. Paradoxically, however, they also gained some rights as individuals, to joint titling with husbands.

From the end of the nineteenth century and for most of the twentieth century, Nicaragua was a highly polarised economy and society. One sector produced coffee, cotton, sugar and beef for export; the rest of society was impoverished. The 'traditional' sector consisted mainly of minifundist holdings with widespread landlessness in the countryside. Within this scenario, the regime of Anastasio Somoza García gained control of this small country in the 1930s. Somoza was followed by his sons, so that the family's increasingly repressive and corrupt regime lasted until 1979. The elder Somoza took power through a successful 1934 plot to murder the peasant guerrilla leader, August Sandino. For Sandino, the land problem was paramount and he advocated cooperativisation and cultivation of crops to feed Nicaraguans, rather than for export. Sandino and his forces' guerrilla attacks became an irritant to the Nicaraguan government, and increasingly to the USA which intervened repeatedly in the early 1930s (Thiesenhusen 1995: 120). Before it withdrew under its 'Good Neighbor' policy (under Franklin D. Roosevelt) the USA ensured that control of the National Guard was placed in the compliant hands of Juan Batista Sacasa, Somoza's uncle. After his uncle assumed the presidency, Somoza succeeded in deposing him. The ensuring Somocista rule was strongman government, such that the political process almost ceased to exist (Thiesenhusen 1995: 122). The Somozas and other elite members acquired more and more of the country's assets. In the countryside, small peasants, including indigenous cultivators, were dispossessed of their holdings as coffee and cotton production became more commercialised.

The Sandinistas and Agrarian Reform

The Sandinistas or *Frente Sandinista de Liberción Nacional* (FSLN) carried out military campaigns in the late 1960s and early 1970s. The eventual revolution in 1979 was achieved at the cost of 35,000 casualties (Thiesenhusen

1995: 127). It was as much a rebellion against a family-based regime as a thoroughgoing social revolution. Significantly, 25 per cent of Sandinista forces were female (Linkogle 2001: 119).

The revolution was greeted by a spontaneous peasant invasion of large estates in the province of León, controlled by the FSLN. The lands were organised into communal farms known as *cooperativas agrícolas Sandinistas* (CAS) or into marketing cooperatives (Powelson and Stock 1987: 240). The Sandinista revolution went on to implement a radical agrarian reform, expropriating nearly half of the country's agricultural land, benefiting over one-third of peasant households (Kay 2001: 759). Somoza-related land was expropriated and reorganised into fifty-three state farms, known as areas of people's property (APP). These were highly profitable export-oriented units that were deemed unwise to divide (Powelson and Stock 1987: 240). Other private lands, however large, were not expropriated if they were efficiently used. Thus, the Sandinistas had a gradualist and pragmatic approach to agricultural policy. They wished to restructure social relations of production in the countryside in an orderly manner, focusing on different sectors at different times (Martinez 1993: 476).

At first, from 1979, the government stressed revival of the agricultural export economy on state farms. Rural workers benefited from wage increases and shorter work days but otherwise saw little benefit (Martinez 1993: 477). Lands already occupied by peasants were declared state land but they were allowed usufruct rights. Other measures included subsidised credit and technical and managerial support for small peasants or cooperatives (Martinez 1993; Ruben et al. 2001).

The agrarian reform policy that eventually emerged from 1981 was a compromise, favouring agricultural production cooperatives (Martinez 1993: 478). Individual ownership, including by peasants, was to be tolerated but would not receive priority.

By the early 1980s, however, the US-backed *contras* had mobilised, and a number of peasants dissatisfied with the collectivist orientation of state policy joined their ranks (Kay 2001: 760). Thus, the eventual shift in policy towards individual household production and land redistribution was provoked by the desire to reduce the influence of the contras, especially given the escalation of counterrevolutionary military attacks on rural targets.[3]

Redistribution of land along individual lines escalated from 1985, when the war against the contras raged. Individual parcels were redistributed more widely, and sharecroppers were given titles to land they worked. Government policy towards cooperatives was also relaxed, permitting much greater diversity including marketing cooperatives. The Ministry of Agricultural Development and Agrarian Reform (MIDINRA) changed its emphasis to encouragement of peasant support for cooperatives rather than assuming that this existed, and peasants in general began to feel the benefits of agricultural policy, including laws that allowed for some expropriation without compensation (Martinez 1993: 481). By 1989, when the agrarian reform was declared 'complete', distribution of arable land was as follows (Martinez 1993: 481):

Percentage of agricultural land by holding

State farms:	11.7
Large capitalist farms:	6.4
Medium-sized capitalist farms:	9.0
Cooperatives:	13.8
Small peasant producers:	49.0

This was a much larger share for smallholders than had been envisaged in 1979, indicating pressure that peasants, or male peasants, were able to exert.

For the economy in general, the Sandinista period was difficult (Colburn 1986; Powelson and Stock 1987: 255). Throughout the 1980s, the country was swamped by inflation (Ruchwarger 1989), counter-revolution and economic malaise (Thiesenhusen 1995: 131). A combination of the contra conflict, labour and market shortages, and conspicuous mismanagement engendered production problems. For instance, there was almost no attention paid to the legal aspects of land titling, fostering insecurity among small producers (Ruben and Masset 2003). A US trade boycott added to the pressures. Coffee production declined by one-third by 1990. Cotton fibre output declined by two-thirds, although cotton producers—with a reputation as highly exploitative employers—received enormous concessions because of the need for export earnings (Colburn 1986; Thiesenhusen 1995: 131). To some extent then, the problems here are indicative of difficulties in constructing 'socialism in one country'.

Against these setbacks should be set the advances made in the areas of health and welfare by the FSLN government. The early 1980s saw a drive to achieve literacy (Arrien 2004). Health services were free, and rural women as well as men also benefited from new housing, electricity, and cleaner water (Collinson 1990: 40).

Although the contra war was not responsible for all problems the Sandinistas faced, it had a large impact. It inflicted devastating losses, with approximately 1 per cent of the population killed and 10 per cent displaced (Kay 2001: 760). The peace deal brokered by President Óscar Arias of Costa Rica traded free elections for an end to outside support for the contras. The FSLN lost the election that followed in the 1990s, gaining only 36 per cent of the rural vote (Kay 2001: 760), much lower than in urban areas. Greater support for peasant smallholders might have yielded different electoral results for the FSLN (Martinez 1993). However, greater engagement might also have held back gender egalitarianism.

Gender Issues and Legislation

Women's emancipation was part of the Sandinista platform, perhaps because of the timing of the revolution after the advent of 1970s second wave feminism, more so than for other Marxist or nationalist-oriented movements. There was a veritable flurry of legislation on personal politics and sexuality in Sandinista Nicaragua, soon after the revolution (Kampwirth 1998). In the first two years of revolution, the feminist influence of the Sandinista Women's Legal Office and the AMPRONAC (women's mobilising organisation) was evident. The first gender law enacted prohibited use of women as sex objects in advertising. Another established penalties for pimping and prostitution, and a third eliminated the distinction between legal marriage and informal unions. An important piece of legislation was the *Ley de Alimentos* ('nurturing law') which declared the equality of all household members. Subsections mandated equal pay for equal work, and that all household members participate in housework and childcare, as well a mandating a maintenance payment from the deserting partner for single or separated women. Part of this initial legislation was the agrarian reform law, unusual in that neither sex nor kinship status (for instance, marriage) was to be an impediment to qualify as an agrarian reform beneficiary. The Agricultural Cooperative Law (1982, Article 132) states explicitly that women should be integrated into cooperatives under the same conditions as men and with the same rights and duties. This reflected both the Sandinista emphasis on gender egalitarianism and the economic situation in which women were active in the agricultural labour force.

A second round of legislation was enacted as part of the 1986 Constitution, which contained ten different articles making reference to women (Kampwirth 1998). This period was one of some struggle within and between women's organisations, including the *Associación de Mujeres Nicaragüenses—Luisa Amanda Espinosa* (AMNLAE), the main Sandinista women's body, more conservative forces, and also independent feminists (Deere and León 2001; Collinson 1990). In the end, provision for recognition of informal marriages or unions was carried but the more contentious 'nurturing' law was dropped as was an even more controversial proposal to legalise abortion (Molyneux 1988).

Feminist organisations came to develop more autonomy and to organise outside the AMNLAE (FSLN) framework after 1987 (Linkogle 2001). Such independent organisation was necessary due to the constraints of the Sandinistas' framework. Sandinista gender politics could be seen as fitting in well with the division between 'feminine' and 'feminist' organising. Feminine organising was around already constituted kinship and gender roles whereas feminist organising asserted rights in furtherance of greater autonomy for women as people, apart from family roles (Alvarez 1990; Linkogle 2001). This is similar to Molyneux's well-known division between practical and strategic gender interests, which itself was formulated in analysis of Sandinista policy (Molyneux 1985).

Given the impact of war, it was difficult to pursue controversial feminist policies without risking further loss of support, especially in the countryside. Whatever was in fact possible, the FSLN accommodated to more conservative forces (Kampwirth 1998), although it did allow much more freedom of organisation than other socialist governments. Despite this, ambivalence existed about women's liberation, as in other Marxist or Marxist-inspired movements. As in examples discussed previously, women were to be liberated by being 'brought into production' and the family was seen as in need of reform to be a more egalitarian unit. At the same time, male prerogatives were not challenged, or not in a wholesale way (Collinson 1990; Disney 2004). Conflicts of interests between the sexes were seen as divisive (Mayoux 1993: 85). Despite this, it seems that feminist organising outside the FSLN and to some extent within trade unions did become more prominent in the late 1980s, perhaps due to the economic and military difficulties the party faced.

Women in Agrarian Reform

The law governing cooperatives specified that women had rights to *involvement* in cooperatives, including at management level, not only rights to membership. Additionally, a 1979 labour law required registration of all workers over fourteen years of age as individuals, so that the wages they received were also as individuals (Mayoux 1993). Thus, the formal legal situation made a serious attempt to legislate for women's participation. Additionally, with greater feminist discussion and agitation in the late 1980s, organisational initiatives evolved. In 1986, the ATC trade union (*Asociación de Trabalhadores del Campo*), the Sandinista Workers' Central, and UNAG (*Unión Nacional de Agricultores y Ganaderos*) all established women's sections. The women's sections held training sessions and discussions on gender issues and helped to establish women's groups in a number of cooperatives (Mayoux 1993: 74).

Women faced a range of difficulties in activating their rights, however, as related by several studies. In general, Collinson (1990) writes that the government did not realise that actions other than legislative ones would be necessary to secure women's new rights. As elsewhere, women household heads were most likely to join cooperatives (see Chapter 4). Previously, most had been landless agricultural workers and cooperatives offered them relative security as well as incomes (Deere 1983). Likewise, many women workers on state farms headed households (Ruchwarger 1989) and women constituted 35 per cent of permanent state farm workers (Deere and León 2001: 161). Many managers did not enter married women separately on payrolls despite their legal obligations to do so (Collinson 1990).

Ruchwarger carried out one of the few studies of gender on state farms in Oscar Turcios, near Estelí (1989). This farm cultivated tobacco for export. Of its permanent members, 60 per cent were women, and of these, 70 per cent were mothers who were single, widowed, or divorced. In terms of gender divisions of labour, men dominated managerial and technical work, including all work with tractors (operating and servicing them), curing tobacco, and security; only a few supervisors were women. Nearly all women worked in production. The rest worked in quality control, as piece-rate monitors or in canteens; 82 per cent worked on the lowest rung of the salary scale. However, as the contra war escalated, more women were able to move up the occupational ladder (Ruchwarger 1989: 90).

Activating gender rights on cooperatives was to prove even more problematic. By 1990, 12 per cent of members were female (Disney 2004), up from 6 to 8 per cent in the previous decade (Mayoux 1993: 73). Most husbands considered it enough that they themselves joined and many opposed wives becoming full members. *Centro de Estudios y Investigaciones de la Reforma Agrária* (CEIRA) studied five marketing cooperatives and eight production cooperatives in the mid-1980s, identifying considerable discrimination against women. Discrimination occurred particularly in access to management positions and technical training. Some production cooperatives excluded women altogether; in reaction women formed production cooperatives themselves. However, other cooperatives had equal numbers of men and women (Mayoux 1993). Women's work was valued most positively where they performed similar work to men, and where they could be seen to be better at some tasks. Formal membership constituted a necessary but not sufficient condition for equal treatment.

Mayoux studied four production cooperatives in Matagalpa and Estelí provinces in the late 1980s, including one woman-only cooperative. In all the mixed cooperatives there had been some increase in women's membership due to drives targeting women. In two cooperatives, women were represented at management level and in general there had been an improvement in attitudes towards women's work. Nevertheless, many women were involved in production without being cooperative members (Mayoux 1993: 81). The women's cooperative owed its existence to state initiatives but had little land per member and no cooperative facilities. Only one cooperative attempted to address issues of childcare, reflecting severe labour shortages. No attempts were made to discuss wider gender issues such as domestic labour or violence.

The micro-politics of a particular collective, El Tule in Rivas province, was explored by Montoya (2003).[4] The study has resonance for the themes of this book, in that it concentrates particularly on aspects of interaction, subjectivities and discourse that help to structure everyday life. El Tule was a 'model' collective with a strong Sandinista identity. Men in the collective were able to hold to 'revolutionary' identities while adhering to their gender privileges, maintaining a form of patriarchal power which was highly controlling of women's sexuality, geographical movement and their social standing. Early on, an underlying ambivalence within the collective surfaced:

> In particular, the collective was plagued by men's relentless attacks on women collective members as *vagas* (vagrants), a term that connotes avoidance of work and . . . [implies] sexual availability. Most men,

including Sandinista militants, accused women of neglecting domestic duties and going to collectives to 'look for men,' some even beating and threatening to leave their wives for participating (Montoya 2003: 62).

In the face of such pressure, most women were forced to capitulate and to leave the collective. Others, however, fought to remain within it.

The ideas of 'good' and 'bad' women and 'the home' versus 'the street' underpinned much gender ideology and much of women's lives; a 'good' woman, who was by definition married, remained at home except in emergencies, confining economic activity to that which could be household-based such as raising small animals. She was modest, faithful, and attentive to her husband and concentrated on performing domestic duties and raising children. Ideally, she was not involved in village affairs and did not gossip, monitor her husband's activities, or question her husband's sexual prerogatives. In return, a man should be able to 'provide' and should protect his family. Village women were subject to high levels of restriction on their movement. After 1979, the building of a school, a clinic and a road widened opportunities for mobility, as did membership of the local AMNLAE. Women who ventured beyond these few prescribed places would risk being seen as being 'of the street' rather than 'of the home'. And the street was—at least symbolically—only territory for 'bad women' as well as for [all] men. As elsewhere, 'the street' functioned as a disciplinary technology to keep women 'in their place' (Montoya 2003).

Despite El Tule's 'model' status as revolutionary, the cooperative was male dominated. Tuleño women established a small, all-female horticultural collective from 1982. After five years the horticultural collective was disbanded in order to form a larger, pig-raising collective. By 1992, however, the number of women in the collective had diminished from 22 to six members. In 1999, the collective disbanded because men from the cooperative demanded their land back. Aside from attempts to deprive the organisation of its land, men constantly criticised women's organisational abilities. Underlying this was the feeling that by working outside the home without absolute economic necessity, women were breaking the terms of the conjugal contract (Pateman 1988). The threat was perceived as sexual and moral. Men often drew parallels between married women collective members and single mothers: 'They walk around like those single women who lack a man's rein.' (*les falta reinda de hombre* [Montoya 2003: 84]). Yet some women persisted in their collective membership despite husbands' opposition. Perhaps symbolically, in 2000, a new, small horticultural collective was established in El Tule on the grounds of the original one.

Struggles also took place within the ATC against gender oppression on state farms (Disney 2004). In the first four years after the revolution, the ATC paid little attention to women's needs. However, in 1983 a large women's conference articulated complaints about direct job discrimination and unequal pay, women's double burden of domestic labour and paid work, and conditions of work for pregnant women (Ruchwarger 1989: 80) as well as childcare and abortion (Disney 2004). Thus, it came to be realised that these issues were not peripheral matters but played integral roles in women's working lives. In the mid-1980s, the ATC began to take action, spurred on by feminist mobilisation within it, agitating for childcare provision on farms and maternity and child sick leave. In 1986 another, much larger consultation took place. One of the issues raised was work norms—the standards of production set for a work day. Above the norm, bonuses could be won—important in a situation of galloping inflation and shortages (Ruchwarger 1989). Women's conclusion was that they wished to have the *same* work norms as did men; however, they required assistance with reproductive labour in order to be able to meet norms (Disney 2004). Their demands included those noted here, as well as communal washing facilities and men's assistance with housework and childcare. Thus, women were keenly aware of the double burden with which they lived, but their choice was to reduce domestic labour in order to be able to participate more fully in the better-recognised state farm sector. In some cases, including the Oscar Turcios state farm, childcare and other facilities were organised, although never on a large enough scale to meet women's needs. Nevertheless, in these cases, women's agitation within a trade union which already had some commitment to equity affected some change.

The Sandinista project reproduced many gender dimensions of the social formation that ostensibly had been rejected. This was done through several means. Firstly, the figure of the 'New Man' was a rather gentlemanly version of an older patriarchal figure and was also posited as the revolution's prime subject (Montoya, 2003). The examples in this volume indicate that this vision was hardly specific to Nicaragua. Secondly, men and women were often segregated into different types of production unit, with women's, predictably, devalued (Disney 2004). Thirdly, patriarchal household structures were confirmed in the late 1980s through the conservative turn in the leadership's stance (Kampwirth 1998; Montoya 2003). Such tendencies were counterbalanced to an extent. The Sandinistas also promoted women's interests, even if incompletely and based on an overly economistic model. There was also some 'space' in this period for grassroots organising, applying pressure for expansion of discussion concerning gender relations, and implementation of women's rights. However, independent feminist voices were always much weaker in rural than in urban areas.

The 1990s and Beyond

Violeta Chamorro was elected in 1990 as the candidate of the opposition (*Unión Nacional Opositora* [UNO]) coalition. As elsewhere, the economic priority was implementation of neoliberal policies. Chamorro also explicitly used conservative and religiously based gender ideology within her campaign, blaming the FSLN for loose sexuality and a high divorce rate (Metoyer 2000), but countervailing forces also existed. In part because of a commitment to civil society shared by Chamorro, feminist voices have gained force, so that women have also made some gains in terms of land titling.

The government had several priorities with regard to land redistribution:

i) resettlement of contra forces as well as Sandinista soldiers, 10 to 15 per cent of contra forces being female (Linkogle 2001);
ii) granting titles to cooperative members and those people with land use rights;
iii) land restitution where it considered land to have been unfairly expropriated; and
iv) privatisation of state farms (Deere and León 2001).

Of 60 former state farms, 30 per cent were returned to former owners; 38 per cent to former soldiers; and 32 per cent to former workers (Thiesenhusen 1995: 136). It should be noted that up to 20 per cent of land was still held in collectives of various types at the turn of the twenty-first century (Ruben et al. 2001: 161).

The issue of titling arose in part because under the Sandinistas up to 70 per cent of redistributed land had never been re-registered, so the Chamorro government faced a mass of claims and counter-claims (Thiesenhusen 1995: 136). From 1993, land titling was under discussion and after six years a large land titling programme was enacted (Deere and León 2001). Despite the partial counter-reform under Chamorro and then Arnoldo Alemán, women made marked gains in terms of land titling (Deere and León 2001). Peasant women in UNAG had begun agitating for joint titles in the late 1980s. Joint titling was made law (Law No. 209) in late 1995, retrospectively covering all beneficiaries of land reform (Deere and León 2001: 205). From 1994 to 2000, women represented one-third of all beneficiaries, representing a considerable advance to their landholding rights during the agrarian reform period in which 10 per cent of direct land reform beneficiaries were female (Ceci 2005). Not surprisingly, local-level difficulties in implementing joint titling became evident early on. In some cases, the provision for *mancomunado* (joint property) was interpreted, for instance, as applying to father-son pairs rather than heterosexual couples.

There is little doubt that joint titling for wives represents a considerable advance. Nevertheless, the impediments women face are often likely to mean that they are unequally positioned in the new market environment. This is one in which smallholder men, too, find it difficult to prosper. The rural poor were hardest hit by austerity measures in agriculture (Davis et al. 2001). Other aspects of neoliberalism such as charges for medical services affected women particularly (Metoyer 2000), and Nicaragua is still one of the poorest Latin American countries (United Nations Children's Fund [UNICEF] 2006).

Increased difficulties faced by smallholders and the poor go hand in hand with concentration of land ownership. Davis et al. (2001) found that micro- and minifundistas controlled 4 per cent per cent of total land although constituting 44 per cent of agricultural households. Medium landowners (holding 20 to 50 *manzanas*) controlled 20 per cent of land (Davis et al. 2001: 172). Smallholders were least likely to be able to subsist on agricultural earnings and sought other income sources. In an interview conducted by Jonathan Steele in 1996, a long-term resident, originally British, commented that ranchers had returned to the area he farmed since the defeat of the Sandinistas. 'They just sort of dropped back, but mightily strengthened: it's the new generation of Somocismo . . . modern, smart, trained in Miami business, and . . . ruthless' (Steele 1996). Thus, women's strengthened property rights exist in a context of increased concentration and of land market sales. These often mean that the poor in practice lose access to land, despite having formal rights.

BRAZIL

Brazil is well-known for both its agrarian movements and for its marked social inequalities. The world's ninth largest economy, in 2005 it was largely an urban country, with over 84 per cent of its population living in cities (Globalis 2008); poverty is less marked in the more developed south than in the north and northeast. Inequalities in landholding figure importantly in the country's social inequities. In 2003, holdings under 10 hectares occupied 1.8 per cent of total land area but constituted 31.6 per cent of farm units (Onsrud et al. 2005: 56). This was little altered from the 1985 figure, when 30 per cent of total units under 10 hectares occupied only 1 per cent of land area (Hall 1990: 206). Properties over 1,000 hectares in size occupied 44 per cent of land area, making up 1.7 per cent of farms. In 2005, Landless Movement figures indicated that approximately five million people were landless or land hungry (Wittman 2005). Unlike the other two cases discussed, Brazil's attempts at land reform have been minimal until very recently.

The land question in Brazil has long historical roots. For 400 years, large plantations, especially in the northeast, were worked by slaves in the main originally imported from Africa. Brazil's economy was oriented to export of commodities, particularly sugar and coffee, and then later, rubber from the Amazon. Modern land ownership was instituted by the 1850 Land Bill. Emancipation of slaves followed in 1888, transferring the economic burdens of subsistence to this new reserve labour army (Martins 2003). Large landowners continued to hold a great deal of power. They were and still are able to influence the police and the judiciary and, often, to rule with violence and impunity (Wright and Wolford 2003).

Three different types of solution to the abolition of slavery developed in different regions. The *colonato* system was introduced mainly in the southern coffee-producing areas, using labour shipped in from Europe partly in order to 'whiten' the population. The colonato was a complex contractual relation, combining salaried labour, payment of rent in labour and goods, and access to plots and animals (Martins 2003: 306). It was eventually superseded by full wage labour. A second response to the abolition of slavery took place largely in the areas of sugarcane cultivation. The mestizo inhabitants of farms (of mixed African and Indian descent) grew their own crops on small plots of land in return for labour services known as *cambão* (the yoke). In some cases, when landowners raised rents, *moradores* (a displaced class of agricultural workers) emerged. These dependent production relations persisted until after the 1950s (Martins 2003: 309). The third 'solution' was based on rubber-tapping in the huge Amazon region. The Amazon has long been employed as a safety valve for dispossessed and unemployed labour, and many rubber-tappers were recruited from the impoverished northeast (Foweraker 1981). Migrants were often trapped in debt bondage through debt to the plantation owner's store (Rocha 2000: 71).

The post-slavery era also saw the development of the important peasant farming sector in the southern states. A minority of families, especially those of German, Italian and Polish origin, were able to become peasant smallholders. Within such households, the male head had absolute control over the labour power of family members (Spindel 1987: 51). Today over four million households are involved in small-scale agriculture, accounting for production of large proportions of basic food items (Onsrud et al.: 82).[5]

A new agrarian phase began in the middle of the century, when cattle ranching began to displace many agricultural workers. From the 1960s, impoverished migrants from other areas of the country could be absorbed in the south-east's industry. For men and women who remained on the large farms, President João Goulart enacted a Rural Worker's Statute in 1963, granting them rights such as inclusion in payroll lists, a minimum wage, a maximum work day, and security after *sixteen* years of work. Goulart also attempted a limited land reform. However, this faltered, as the USA cut off Alliance for Progress funds (Lapp 2004).

A military regime displaced the moderate Goulart in 1964, ruling until 1985. Aside from political repression, it pursued economic strategies favouring large corporations; in agriculture, mechanisation was adopted and corporate agricultural interests encouraged. In sugar production, plantation owners increased rents and many tenants were evicted; the populations of *favelas* (shanty towns) soared. Corporate lumbering and cattle ranching interests also received priority.

The ensuing crisis for the poor entailed by these policies resulted in some direct state encouragement of migration to Amazonia in the early 1970s.

This substitute for land reform failed for a variety of reasons, including the Amazon's poor soils (Branford and Glock 1985) and lack of organisation and of support for settlers. Only 8,000 of a projected 100,000 families were resettled in this way (Hall 1990: 208). Other programmes were enacted in the western states of Rondônia and Acre. Again, these were beset by poor project design and support. Most migration to the Amazon, however, was not state sponsored but was *ad hoc* and informal: an attempt to escape severe poverty and unemployment. In the Amazon even more than the rest of the country, few formal titles existed, including by those operating rubber estates (see Branford and Glock 1985). Many Indians and peasant squatters were 'legally' evicted from lands. In the frontier regions of Pará, the Mato Grosso, and Paraná, violence has been consistently exerted against both Indians and peasants by large landowners or others asserting land rights (Foweraker 1981). Companies, the police, hired gunmen and, sometimes, the military operated together.

This brief outline of Brazil's complex agrarian social history serves to indicate the pressing need for land reform for much of the twentieth century. By the end of the twenty-year period of military rule, 1 per cent of farms encompassed 44 per cent of land and the Gini coefficient of inequality was 0.853, one of the highest in the world (Deere 2003: 262). It has been in part through actions of rural and peasant movements that reforms to this structure have taken place in Brazil.

Resistance Movements

Movements based on smallholding agriculture have a long Brazilian lineage. The first recorded resistance concerned runaway slave communities which managed to remain independent for most of the seventeenth century under *Zumbi de Palmares*, an iconic figure among black communities.

In the twentieth century it is likely that most resistance by diverse movements has gone unrecorded. The Peasant Leagues of the northeast are some of the most well known. They were first established in Pernambuco state in the 1940s, with modest aims, but in a few cases, the Leagues campaigned for land reform. Redistribution of at least one estate was accomplished in 1959 (Wright and Wolford 2003: 129). These organisations spread rapidly from the late 1950s, and by 1964, 2,000 Leagues existed in twenty-four states. The considerable advance represented by the 1963 Rural Workers' Statute was partly a response to the Leagues, which were violently suppressed by the military government.

In the early 1970s, peasant guerrilla movements formed against police, *gato*, and landlord violence, but despite much 'counter'-publicity, they were very small (Foweraker 1981: 23). Another movement, *Movimento dos Agriculturas sem Terra* (MASTER) campaigned for agrarian reform in Rio Grande do Sul in the south. It was begun by children of European immigrants who had failed to gain land to cultivate (Onsrud et al. 2005). A third movement grew up in response to the regime's policies of construction of large hydro-electric dams, flooding the land of small farmers, and facilitating agro-industry; in this case, in soybean production (Rothman and Oliver 1999).[6]

In response to the 'need' for migration to the Amazon, the government established National Institute for Rural Settlement and Agrarian Reform (*Instituto Nacional de Colonização e Reforma Agrária*; henceforth INCRA) in 1971. In the early period, INCRA's duties were limited to assisting *asentamento* (resettlement) of those displaced by the dams and in government-sponsored colonisation schemes. Perhaps improbably, a significant piece of legislation was also passed in this era: the 1964 Land Statute established that suitable land not serving a 'social purpose' or not productively cultivated, could be expropriated for agrarian reform.

During the two decades of military regime, liberation theology was taking root. Socially minded priests and nuns began to organise *Comunidades ecclesial de base* (CEB; Christian base communities). This wing of the Roman Catholic Church saw inequitable landholding as a fundamental factor in social injustice. Thus a Pastoral Land Commission was established in 1975 to support peasant land struggles in the Amazon. It aimed to counter landlord abuses, especially the massacre of peasant leaders and frequent evictions and land-grabbing (Adriance 1995). The CEB came to be prominent in rural unions and in land occupations (Adriance 1995) and helped to provide legal counsel.

One of the few studies discussing women's agency in the agrarian social movements of the period concerns the upper Amazonian CEB (Adriance 1995). A number of women were active in land occupations and some became leaders; more women than men participated in the CEB Adriance studied (1995: 149). As in the other Latin American cases discussed, women often had conflicts with husbands over their participation, although participation was facilitated when the men were already CEB members. The husbands of women activists were more likely to stay at home taking care of children than were other men. Women in the CEB studied were, despite their participation, primarily oriented to their family roles and placed a high value on fertility, in line with religious teachings. Nevertheless, participation had developed their belief in their right to a life outside the home.

From the late 1970s on, the military government began to adopt a policy of *abertura* (opening up), and various social movements were able to emerge. However, the first completely democratic elections (with universal franchise) were held only in 1986.

In western Amazonia, rubber tappers' unions were formed, with some support from the CEB; women as well as men were active in the unions (Campbell 1996). These aimed to protect tappers' livelihoods as well as the forest environment. Eventually, from 1987, they were successful in establishing protective reserves similar to land reform areas. In 1984, the *Movimento dos Trabalhadores Rurais sem Terra* (MST; Landless Movement) was formed, originally as an offshoot of the CEB. The MST has been the most successful agrarian reform movement to date in Brazil, and is, arguably, the best known globally; it will be discussed further. A counter-reform movement of landowners, the Union for Rural Democracy (UDR) was also formed.

The abertura also opened the way for explicitly feminist movements in Brazil. These have been largely urban but have sometimes had an impact in rural areas, partly through influencing other social movements. It is usual for rural women's movements to be weak but in Brazil an additional reason exists: unusually, agriculture is *not* becoming feminised, or at least not at the same rate as elsewhere, reflecting high rates of female out-migration from rural areas and young women's desire to have urban jobs (Guivant 2003; Wright and Wolford 2003; Brumer 2004).

In the southern states, an umbrella rural women's movement, the Rural Women Workers' Movement (*Movimento de Mulheres Trabalhadores Rurais*, or MMTR) was nevertheless formed in 1989, developing from activities of women working within other movements such as land reform movements, anti-dam movements, the CEB, and rural labour unions. By 1992 it had over 30,000 members (Stephen 1997: 209). By the late 1990s, the MMTR's agenda had shifted, from concentration on working conditions, benefits, and land issues to include specifically feminist issues such as reproductive health rights, women's political representation, and domestic violence. Although MMTR leaders had had much contact with (and experience in) other social movements, they had often been marginalised when raising gender issues. It is significant that MMTR leaders have been influenced by feminists in other movements and in NGOs. The MMTR maintains links with the church, CEB, the landless movement, and with trade unions but remains autonomous. It is one of the arguments of this book, that without autonomous women's organising it is unlikely that women will be able to achieve equity within social policies such as agrarian reform, even though they are meant to be transformative.

Land Reform Strategies and the Market-Led Reform

In the early and mid-1980s, a campaign for a National Agrarian Reform Plan (*Plano Nacional de Reforma Agrária*, or PNRA) took place, and in 1985 the PNRA was passed by government (Spindel 1987: 61). Following this, a new constitution in 1988 gave government the right to expropriate unused land for 'social use'. Core demands of the PNRA were largely defeated by the land-lords' union with only expropriation of unproductive latifundia being allowed. Landowners were also allowed to keep 25 per cent of land and to choose which parcels to keep (Deere and León 2001; Wright and Wolford 2003).

Throughout the 1980s and 1990s, the MST as well as other groups mobilised for land occupations and were often met with great violence. These initiatives provoked two state reactions. One response was increased land expropriation and settlement on state schemes. These, however, were limited; between 1964 and 1994, only 850 land reform settlements had been created, benefiting approximately 144,000 (Deere 2003: 277).

Additionally, the government instituted a market-led reform (Market-led agrarian reform, or MLAR) under World Bank auspices. As elsewhere, the MLAR strategy tries to circumvent state-directed land reform. It seeks cooperation of landlords ('willing sellers') who will sell to peasants or other 'willing buyers'. Buyers must form organisations with a 'business plan' to achieve economies of scale and also in order to weed out inexperienced or unsuitable applicants (Borras 2003: 373).

In practice, the MLAR has several flaws. Perhaps intentionally, MLAR beneficiaries have above average incomes; elite peasant leaders have taken control over the process rather than this being enacted from below (Borras 2003). Relatedly, land prices have been much higher than in state-led programmes, although one of the initial intentions was to avoid landlords setting prices over market rates (Rocha 2000: 22; Borras 2003). Another issue is that the extension service has been privatised and is of poor quality (Borras 2003: 388). Additionally, settlers have often been overwhelmed with commercial debt (Rocha 2000). All of these factors have serious implications for gender equity: in order to farm independently, women require a good quality extension service and preferably one which includes female staff. Peasant women are even more likely than poor peasant men to be vulnerable to dispossession of any newly acquired land in case of debt (Manji 2006). Borras concludes that the MLAR in Brazil as in other countries has failed even on its own terms.

In January, 2003, Luiz Inácio Lula da Silva (known as 'Lula'), an ex-trade unionist, was elected president and was re-elected in 2006. Despite hopes of a more extensive land reform under his presidency, a dual path has been steered, continuing both to support agribusiness and tolerating land occupations and pledging support for agrarian reform (Welch 2006). In 2006, a second PNRA (II PNRA) was promulgated, combining elements of market-assisted and traditional or state-backed land reform. This states the intention to support small-scale agriculture, including special lines of credit such as one for women (Onsrud et al. 2005: 86).

Gender and Land Reform

The general trajectory of land reform in Brazil, then, has been halting and fragmented. Due to the weight of legal and social custom, it has been difficult for women to benefit from land reform and property, including land, is heavily concentrated in male hands (Barsted 2005). Within this frame, there have recently been some attempts to include gender equity within various programmes and policies of the state and social movements. The 1988 Constitution established an explicit right for women to obtain land tenure concessions or titles within agrarian reform and guarantees equality between husband and wife. However, in practice gender issues have not been recognised fully in agrarian reforms, and Brazil lags behind in terms of application of gender progressive statutes (Deere and León 2001). INCRA uses a points system to select beneficiaries: until 1988 it was legal to award men one point, but 0.75 points for women and children. By the mid-1990s, only 12.6 per cent of agrarian reform beneficiaries were female (Deere 2003; Guivant 2003). The criterion that beneficiaries should be experienced in agriculture tends to disadvantage all women, since much of their work goes unrecognised. Another impediment is that registration of names of both spouses on documents is optional in Brazil, so in most cases the husband continues to represent the household. Underlining this, the forms used by INCRA for many years had room for only one beneficiary, with the wife usually listed as first dependent, on a separate page (Guivant 2003). Because of pressure from feminists, a redesigned form now includes room for the wife's signature on the first page.

The *Movimento dos Trabalhadores Rurais sem Terra*

The MST has evolved into an organisation whose membership is 1.5 million with a presence in 23 of 27 states (MST 2006). In 2001, it claimed 1,600 settlements with 250,000 families as well as 500 'encampments' with 70,000 families. It has launched numerous land occupations and has pushed land reform onto the national agenda. The MST's aims are broader than simply land reform, with roots simultaneously in Marxist and populist thought, in liberation theology, and also with influences from Paulo Freire concerning cultural change. It sees itself as a revolutionary and transformative movement seeking to create 'new' men and women, to alter Brazil's landholding system and Brazilian social structure more generally. It has added aims such as promoting sustainable and organic agriculture. In 2007 an MST Congress decided to prioritise ecological questions (Caldeira 2008).

The MST's main tactic to acquire land has been to organise mass occupations of estates considered as possibilities for expropriation, often because the land is not fully utilised. Occupations are often dangerous, with many militants having been injured and a number murdered. Following the occupation, those involved must withdraw and wait in temporary encampments, often by roadsides and in makeshift shelters while INCRA assesses the land occupied to determine its suitability; the law states that occupied land will not be redistributed. This period of waiting often takes a number of years as the legal process proceeds (Caldeira 2006). However, encampments are organised as quickly as possible, and schools, health posts, sanitation, and basic food rations are provided. Only when the land has been acquired by the state can a more permanent 'settlement' be established.

The MST's remarkable achievement is to cement land and agrarian reform in the national (and regional) political agenda(s). Beyond this, it is due largely to the MST's public profile that landlords are now far less free to dominate affairs in their regions, including the appointment of public officials and of judges and the abduction, eviction, torture, and murder of the rural poor, although this has not disappeared (Navarro 2005: 133).

The MST has also successfully promoted functioning land reform settlements or communities. It has been attentive to provision of services such as schools, clinics, marketing of produce, and (in some cases) encourages cultivation of crops using organic methods (Onsrud et al. 2005), initiatives that benefit all settlers, and perhaps particularly, women. A tension, however, is that whereas leaders prefer collective models of production, settlers often prefer individual household production (Caldeira 2006); for this reason, collective production has been dropped as a national model (Welch 2006: 207).

Organisationally, the MST is meant to be democratic and participatory: all adult family members must join rather than simply the household head. Many meetings are held both in encampments and settlements and participation of ordinary settlers is sought (Onsrud et al. 2005). The MST makes some attempt to avoid the patronage relations endemic in Brazil through mechanisms such as drawing leaders from distant regions and limiting their stay in particular settlements (Caldeira 2006). Some, however, doubt that leaders are openly selected (Navarro 2005: 135).

The MST has had an ambiguous relation to questions of gender. A National Commission of Women (CNM) was first established in the mid-1980s, only to be disbanded in 1989. In 1996, the CNM was re-established (Guivant 2003). In 2001, the CNM, by then a constituent body of the MST, put forward a statement calling for discussion of women's roles in the movement (Giani et al. 2003). In 2006, the MST's website affirmed its commitment to gender equity, recognising that gender discrimination was a current issue. Its specific aims in 2006 included:

- a guarantee of childcare to aid women's participation at work and in meetings;
- 50 per cent of women in all training and activities;
- joint land titles for spouses;
- one male and one female coordinator to live in base communities; and
- prioritising campaigns to provide documents for women workers.

Women have been very prominent in MST land occupations (Wright and Wolford 2003), as the organisation recognises. Women are also well represented in central leadership, with nine of 23 leaders in the highest echelon being female (Harnecker 2002: 14).

The MST, however, also sees the basic social unit as the nuclear family—indicative of its religious roots in the CEB as well as of tendencies common to populist movements (see Chapter 2). Harnecker is optimistic about the possibilities of the model, seeing the MST as putting forward a smallholder

Figure 7.1   Logo of the *Movimento dos Trabalhadores Rurais sem Terra* (MST), Brazil.

democracy in which peasant patriarchy has been overcome. Others are less hopeful, noting that the MST tends to see gender and class analyses as contradictory (Razavi 2006: 2). In interviews with movement leaders, Caldeira (2008) found that some considered gender a 'distraction' from the class struggle (2006). Following the 2007 Congress, the website section on gender was removed, although mention of domestic violence remains (MST 2009). Thus, the space given for discussion of gender discrimination within the movement appears to be decreasing.

These events indicate deep ambivalence concerning gender, including whether it should be taken up by the movement as a serious issue. Such ambivalence is rooted in the importance of the nuclear family model for the MST's vision. This model is contradictory for women in at least three ways. The first two concern the invisibility of women's work and the possibilities for male control within the household unit. The third is that many households are not nuclear but are female headed. In selecting beneficiaries, INCRA gives priority to larger families, which in itself disadvantages lone-parent households. Female household heads constituted over 12 per cent of rural households in 1996, but only 5 per cent of land reform beneficiaries (Deere 2003: 270). In encampments the MST's rules inadvertently disadvantage female-headed households by stipulating that at least one member of the household must remain on the land. Thus, female household heads cannot leave to find waged work. A Rio state study showed that many were unable to sustain this situation over a long period and ended up abandoning encampments despite participation in the initial struggles (Caldeira 2006). When a lone woman gets married, her land rights may be transferred to the new husband (Deere 2003).

A number of sources acknowledge that women are active in land occupations, confrontations with the police, and in the encampments; however, once land is gained, they are often expected to return 'home' and to their small plots (Deere 2003). This parallels women's experiences in many nationalist movements (Jayawardena 1986; Yuval-Davis 1997). Several studies of gender within MST settlements indicate that inequitable treatment continues along with advances such as the achievement of food security (Onsrud et al. 2005). For example, a study by da Silva of an MST settlement in Santa Catarina state stresses lack of attention given to women's domestic burdens and the deep-rooted nature of the gender division of labour. Da Silva says (2004) that home is seen as very important by many women, as they may have been homeless previously. Women are usually part-time members of the cooperative whereas men are full-time. Accordingly, women often receive half the level of male payments. In general, men do heavy work such as driving tractors and women do lighter work. This study also found ideas about the nuclear family to be highly significant, in that the MST council intervened in the sphere of sexual morality. Adultery is considered a transgression against the collective and to maintenance of good relations within it. Da Silva cites an incident of adultery in which the woman involved was found 'guilty' and advised to leave the settlement, whereas the man involved stayed. Moreover, the man was a leader so that extra-domestic power relations were involved. Thus, the gap between rhetoric and practice in this case was considerable.

Recent Initiatives

In recent years, especially since 2000, Brazilian rural social movements have attended more seriously to gender issues in a number of ways. One development is in the rural unions which may have been influenced by movements such as the MMTR. The largest rural trade union, *Confederação Nacional dos Trabalhadores na Agricultura* (CONTAG), as well as the left union

Central Union of Workers (CUT) have been important both in campaigns for land rights and for benefits for all workers, including women. However, the unions do not focus on land reform as such. The CUT has been particularly important in pushing gender onto the agenda, leading to the establishment of a pioneering National Committee on the Question of Women Workers and to formulation of a gender policy (Barsted 2005: 48). More female union leaders have emerged in recent years. Gender discrimination certainly exists, but the CUT has established a minimum quota of 30 per cent women in leadership positions, although it struggles to fulfil this (Boni 2004).

The rural unions have come to be more proactive in effecting change on gender issues. CONTAG, for instance, led the march of the *Margaridas* in 2000, the largest-ever demonstration of rural women with approximately 20,000 participating. The women put forward both general demands concerning land reform and gender-specific ones. The latter included an end to sexist violence in rural areas and campaigns for access to education, to credit, and for reproductive health. They also criticised the low percentage of women beneficiaries of agrarian reform (Guivant 2003: 29).

Another notable development is that of the formation of a national peasant women's movement, the *Movimento Mulheres Campesinas* (MMC), established in 2003 with representatives from fourteen states, including women from church organisations, the MST, the anti-dam movement (*Movimento dos Atingidas por Barragens*, or MAB), rural labour unions, and other organisations. The MMC's website states that it is socialist as well as feminist. Agrarian reform together with calls for environmental protection are among its aims. This is a hopeful sign, and it is unusual for an autonomous organisation to exist. However, the difficulties of raising feminist issues within the MST, the most prominent agrarian reform movement, remain.

CONCLUSION

These three Latin American cases differ in the extent, duration, and trajectory of agrarian reforms. However, they also display continuities. In general in Latin America, women have been highly marginalised within land and agrarian reforms. Where cooperative membership is important, as has frequently been the case, married women are rarely members in their own right. Divorced single and widowed women with children (that is, 'household heads') may be able to obtain or to inherit land or membership rights. However, even this is not automatic. For married couples, the husband's membership has been assumed to be 'for' the family, and therefore, women are marginalised to the point of being excluded. Even when this is not the case and women have individual membership rights, as in Nicaragua and MST settlements in Brazil, a range of informal mechanisms or processes serve to keep them marginalised.

Such processes operate through gender regimes, which although subject to flux and transformations, also may display a good deal of continuity. The cases discussed here all concern *mestizo* or majority and Spanish-speaking populations, and indigenous communities may have different and slightly more egalitarian models. However, the strength of ideas of domesticity and of marianismo for women is striking in its model of women as mothers and domestic workers, semi-confined to the home and with attributes of patience, modesty, and endurance. This gender regime is buttressed by machismo. Examples given here appear to indicate that these are not simply outmoded stereotypes, or at least, not in the countryside. They constrain women's actions even in the economic sphere and limit their room for manoeuvre, while extending that of men as husbands.

In order to combat such beliefs about women's 'proper place' and men's proprietorship over women, a government, state, or social movement would need to go beyond productionism. The Engels-derived model in Latin America as elsewhere (see Chapter 4) has made some gains for women, particularly for widows and divorcees who sometimes acquire or else inherit land rights or cooperative membership. That most left governments and movements have a basic commitment to improving women's legal and economic status is also a gain that should not be derided. However, the productionist model has serious limitations, particularly in the absence of attention to women's domestic labour. Failure to attend to ways of reducing women's work burdens has often simply resulted in increased workloads. Even when movements are cognisant of the extent of obstacles facing peasant women, and wish to ameliorate these, this perspective is usually counterbalanced by worries about the unsettling affect on peasant men and loss of their support. Thus, women may gain rights at the formal level but these often remain unenforced or else partially enforced, as in the Nicaraguan agrarian reform.

The examples given offer some indication that the practice of family agriculture is integrally bound up with household models of land reform, as this book argues. It is far from a simple matter to integrate women's rights into such a model, and it is for this reason that the 'woman issue' is often ignored. It seems that trade unions find it easier in the current juncture than land movements to incorporate women's demands, as indicated in the Nicaraguan and Brazilian cases. However, they do not have agrarian reform as a main focus.

Lastly, these examples begin to indicate the difficulty women have in enforcing and activating rights to land and property, even within favourable legal environments. It is to be hoped that the autonomous rural women's movements now forming in countries such as Brazil will make gender issues more visible within rural movements.

8 Land Reforms, Customary Law, and Land Titling in Sub-Saharan Africa

The question of gender and land reform in sub-Saharan Africa is framed by competing discourses and policy processes. One discussion concerns the ongoing food security crises on the continent. Another is the familiar theme of market-based versus state-backed redistributionist reforms. A third concerns individual versus collective land titles. In African contexts, the 'collective' usually refers to land administered under traditional authorities in lineage-based systems rather than to collective farms (see Chapters 4 to 6).[1] The present chapter concentrates on the southern African case studies of Zimbabwe and South Africa. The dramatic example of Zimbabwe indicates some of the pitfalls as well as possibilities of land reform within what is to date, the largest scheme in Africa. The chapter also explores the conundrums thrown up in contemporary debates over gender, land, and land titling elsewhere, using the examples of Tanzania and Mozambique.

CUSTOMARY LAW

Gender and land issues in sub-Saharan Africa are framed by customary law, and so the chapter begins with an examination of this. The land tenure systems of most rural people on the continent have been governed by customary law and subject to chiefly authority. Land is crucial to the unity of lineages: 'Land helps to order relationships between people, living and dead' (Davison 1988a: 15). Under customary law, land was held on behalf of the lineage and its ancestors by elders or a chief and could not be bought or sold. Thus, the advent of land and agrarian reform programmes in Africa encounter circumstances different to those of the privatised systems predominant in Latin America and Asia. Central to these circumstances is the place of women within customary law. In patrilineal systems, land is held by men as lineage descendants. Women access land through husbands and land rights are usually inherited by elder sons.

The term 'customary law' was a colonial (British) invention (Ranger 1985). In the system of indirect colonial rule, there was an attempt to retain tribal or lineage practices 'intact', although the wider economic policies of colonialism in fact undermined the economic bases of traditional systems. 'African custom and tradition' was allowed to regulate certain matters such as landholding, marriage and family law, and restitution for petty crimes, with cases arising from customary law usually heard by primary or lower courts (Banda 2005). In ex-settler colonies such as Zimbabwe, South Africa, or Kenya, African people were relegated to poor quality land in remote areas in 'reserves'. In west Africa, customary systems of land tenure hold sway more generally in rural areas which lacked significant settler presence.

Much debate exists about the place of women in customary systems in pre-colonial times. Some argue that customary law gave women a number of rights (Hellum 2007), and many agree that it operated as a more flexible and locally specific system in the past. With colonialism, customary law became codified and a flexible system became ossified, with detrimental effects for women. Chanock has argued that the codification of customary law reflected an informal consensus between African lineage elders and colonial officials, as both were convinced of female inferiority (Chanock 1989). Women became 'legal minors', unable to hold land on their own behalf, unable to sign or agree to contracts (including marriage contracts) without a male representative, unable to represent themselves in court, or to retain custody of children after early childhood. Although these practices and others such as bridewealth payments and child betrothal were practiced in pre-colonial times, it is sometimes forgotten that younger men as well as women were 'minors'. Because widows do not have rights to patrilineage land, customarily they were expected either to return to their natal homes or to retain access through a levirate marriage with the deceased husband's brother. Patrilineal systems, which form the majority,[2] are also potentially polygynous and it is much easier for men than women to obtain divorces. For instance in contemporary Uganda, a man has only to accuse his wife of adultery to obtain a divorce (ActionAid 2005: 62). In Zimbabwe, a woman's refusal of sex or use of contraceptives is ground for divorce, as is 'insubordination' or vocal disagreement with the husband.

However customary systems operated in the past, it seems clear that current interpretations disadvantage women. For instance, customarily wives had the right to a plot of land to cultivate 'women's' or food crops. Although this land was accessed through the husband, virtually all married women would have had access. With the growth of a cash economy and with land shortages, the granting of this plot became much more discretionary. Similarly, in Zimbabwean Shona society, women traditionally had the right to profits from *mavoko* ('of her own hands') activities, such as midwifery, beer-brewing, traditional medicine, or vegetable crop sales. In contemporary circumstances these rights have been eroded. Many husbands—even men with wives in formal, urban employment—appropriate women's incomes, claiming that payment of *lobola* (bridewealth) gives them this right (Jacobs and Howard 1987). The advent of cash economies and migration meant that checks on behaviour from the community were eroded and individual men gained more power over wives (Pankhurst and Jacobs 1988; Busingye 2002).

Customary law is increasingly threatened from a combination of forces such as more gender egalitarian statutory law and marketisation. However, customary law practices continue to have profound effects on women's land rights across the continent (Wanyeki 2003), despite both variation and change. With regard to gender and land rights, fierce debate exists about the implications of customary versus individualised tenure. Feminist lawyers have found themselves pitted against land rights advocates in policy debates.

Land reform programmes are usually viewed as aspects of general 'modernisation' schemes and so have sometimes presented opportunities to reconfigure customary law with respect to women's landholding rights. For instance, widows have gained some rights as many are left destitute when the husband's relatives claim her property. Women's land and other rights *within* marriage have, however, proved much more contentious.

ZIMBABWE

The Zimbabwean crisis has brought the question of agrarian reform to unusual prominence. At the time of writing, the country's situation is chaotic following a heavily disputed run-off election for president in June 2008 amid accusations of vote-rigging, fraud, and great violence against Movement for Democratic Change (MDC) supporters. Inflation stood at 231 million per cent in July 2008 (British Broadcasting Company [BBC] 2008); 94 per cent were unemployed in 2009 (McGreal 2009). Life expectancy in 2006 was the lowest in the world, at thirty-four years for women and thirty-seven for men (World Health Organization [WHO] 2006). Shortages of most goods are severe. In 2006 to 2007, 4 million people required food aid (McGreal and McVeigh 2007); in early 2009 the United Nations (UN) estimated that 7 million of Zimbabwe's approximately 12 to 13 million population required food aid (McGreal 2009). Shortages of food and other goods are to some extent eased by the fact that at least one-fourth of the population has migrated abroad due to the political and economic circumstances faced, mainly to neighbouring South Africa and the United Kingdom.

The scenario of economic meltdown is usually linked with the 'fast track' land reform initiated in February 2000, when the government began expropriation of white-owned commercial farms. Thus, land reform has come to be seen as the root cause of the crisis. Here I argue that the problem has been the political manipulation of land and land redistribution, rather than agrarian reform *per se*. In the current situation it is easy to forget that a well-managed and relatively successful land reform took place in Zimbabwe prior to 2000 and that other scenarios for land reform are possible (Kinsey 2004b). It is also easily forgotten that the great majority of the rural population, up to 70 per cent of the agricultural labour force, is female (Permanent Mission of the Republic of Zimbabwe 2000).

This section of the chapter firstly discusses land reform in Zimbabwe prior to 2000 and secondly, its gendered impacts. It then examines the reasons for the current fast track land reform and gives an overview of its results. The last sub-sections consider fast track land reform and its gendered implications.

Land and Land Reform pre-2000

Inequality in landholding is deeply racialised and has indeed been a root cause of inequity in Zimbabwe in the past as well as the present. White control over land has been crucial in securing economic and political dominance for Europeans. British-origin and other European settlers marched north from South Africa in the 1890s, hoping to find the 'Second Rand' of precious metals and minerals. For a time the settlers' search set a premium on food production, and Shona peasants in east and central areas farmed more intensively to produce food for sale (Ranger 1985). When the territory which became Southern Rhodesia was found to be poor in minerals, settler attention turned to the development of large capitalist farming estates. Land expropriation served to ensure Europeans' need for land. A series of taxes were imposed on African men to ensure that they entered the cash economy, and they turned to wage labour in mines, in European households, and in agriculture in order to meet 'head taxes', 'hut taxes', and other extraneous levies, including a tax on dogs (Arrighi and Saul 1973; Palmer 1977). By 1930 with the Land Apportionment Act, 51 per cent of land was declared to be European and 30 per cent reserved for Africans (Herbst 1991). The reserves were to continue under communal or chiefly tenure. Various other pieces of discriminatory legislation undermined African peasant farming. For instance, the Maize Control Act of 1931 subsidised European maize production while reducing prices African farmers could command for their main marketed crop.

Agricultural land in Zimbabwe is variable in terms of soil quality, and only some 17 per cent of land can support intensive rain-fed agriculture and livestock production (Weiner 1988). Land was, and is, classified into natural regions according to rainfall and soil fertility. Natural Region I (located mainly in the east) is of high fertility and relatively reliable rainfall while Natural Region IV is suitable only for extensive livestock production. Europeans appropriated not only the majority of land but the most fertile, with large farming estates being concentrated in Natural Regions I, IIa, and IIb. African reserves were located mainly in Natural Regions IIb, III, and IV. Already arid, these became overcrowded, and soil quality deteriorated further. Men who were able to, migrated to cities and mines to work, while women, elderly men and children stayed behind to farm. Women undertook ever increasing shares of work although agriculture was becoming less viable.

In the 1950s, several government measures attempted to reform the migrant labour system and to form a more viable and permanent African peasantry and working class. Eight per cent of land was set aside for Africans to buy on a freehold basis in Native Purchase Areas. The Native Land Husbandry Act (NLHA) of 1951 attempted to replace customary tenure with land titles. However, this increased landlessness and rural inequality (Phimister 1983)—the only way to designate the landed entailed designation of the landless. The NLHA was in any case abandoned with the election of the Rhodesian Front in 1962.

The Unilateral Declaration of Independence (UDI) from Britain was declared in 1965. The guerrilla war temporarily uniting the Zimbabwe African National Union (ZANU) and the Zimbabwe African People's Union (ZAPU), under Joshua Nkomo, began shortly after UDI. One of the main aims of the guerrilla struggle was the return of 'lost lands' (Ranger 1985), others being improved livelihoods, an end to racial oppression, and the wish for more democracy in daily life. A negotiated agreement known as the Lancaster House Agreement ended the war in the late 1970s and led to free elections.

Land resettlement 1980 to 2000

ZANU, famously, won an electoral victory under Robert Mugabe in 1980. One of its first priorities was to establish an agrarian or land reform programme. At independence, most prime farmland and nearly all large farms were owned by whites. Approximately 6,000 large-scale commercial farms (LSCFs) and a number of corporate agro-industrial estates occupied 39 per cent of Zimbabwe's total land area and two-thirds of the most fertile areas (Takavarasha 1998: 2). They also produced over 90 per cent of marketed output (Weiner 1988: 74).

The aim of land reform (termed 'land resettlement') was to redistribute land to the poor and landless or near-landless, and additionally to establish a class of self-sufficient or better off African smallholders. From the outset, the terms of the land reform programme were set by the Lancaster House Constitution. One of its provisions, in force until 1990, was that land had to be acquired on a 'willing buyer–willing seller' basis. Land was distributed under several programmes or 'models' including a small number of state farms and production cooperatives (Model B). Large numbers of people also settled land informally (by 'squatting') following independence and thereafter (Chimhowu and Hulme 2006). Most land distributed formally was under the Model A scheme to individual households. Under Model A, households were allocated five hectares or approximately 12 acres of land. This was held via a series of permits by the household head.

Variable amounts of land were allocated for cattle grazing, with more allocated in the more arid areas of Natural Regions III and IV. Another feature of Zimbabwe's land reform was that settlers were meant to be full-time agriculturalists. In 1980, beneficiaries were required to be age 18 to 55 with dependent families, married, or widowed and not in formal employment (Government of Zimbabwe 1985). The stated intention of the programme was to resettle the rural poor rather than the better off (Stoneman 2000), although from the late 1980s increased attention was given to raising productivity (Kinsey 2004b).

Resettlement Areas (RAs) were conceived as areas separate from the Communal Areas (CAs). Until early 2000, the RAs were administered by appointed Resettlement Officers (ROs) who had quasi-judicial powers. The policy was to administer individual families rather than whole communities (Kinsey 2004b). This meant that many were likely to be living with new and unfamiliar neighbours and also implies that many people found themselves living on lands believed to belong to the ancestral spirits of others' lineages (Jacobs 1989).

Early targets for numbers of households to be resettled were very large, with the aim of redistributing land to 162,000 households by 1984 (Agricultural Extension Service of the Government of Zimbabwe [AGRITEX] 1985: 15). By 1985, 38,000 households had been resettled, 35,000 in Model A schemes (Hanlon 1986: 8). By early 2000, 91,000 households had been resettled, of which 61,380 were in Model A villages (Chimhowu and Hulme 2006: 731). Thus, over half of resettlement took place in the first years after independence. This was also the period in which land was easiest to acquire as people could be resettled on underutilised or deserted farms (Cusworth 2000). Nevertheless, Zimbabwe's land reform has been the largest attempted in Africa (Kinsey 1982).

Land reform has been of importance since independence at the rhetorical level, tending to 'surface' particularly at election times (Alexander 1994). However, commitment in practice has been more variable. For a period from 1985 on, official attention shifted to raising productivity in the CAs where the majority of the population lives. This was attempted through provision of credit, irrigation, and agricultural extension services and was broadly successful (Zimyana 1995) while state support was available. From the late 1980s, with the expiry of the Lancaster House Constitution imminent, land re-emerged as a more important issue (Palmer 1990; Alexander 1994). In 1990 President Mugabe stated that 50 per cent of commercial farmland might be purchased for resettlement. A Land Acquisition Act was passed in 1992 to facilitate such a process.

In the early 1990s, however, a structural adjustment programme (ESAP) was launched. As elsewhere (Mohan et al. 2000), this led to deindustrialisation, reduction of the public sector, and severe drops in wage levels. Government expenditure was directed to maintaining health and education sectors as a higher priority than resettlement (Moyo 1995). However, structural adjustment and a neoliberal orientation benefited commercial farmers who produced export crops such as tobacco, speciality vegetables, and ostrich meat (Moyo and Yeros 2005). By this time, too, the full impact of the HIV/ AIDS epidemic was evident; between 1995 and 1997, HIV/AIDS infections peaked at 29 per cent of the adult population (AVERT 2008a). Land reform receded in terms of importance in policy.

Much debate exists concerning the ZANU government's commitment to agrarian reform from the 1990s. Here, I give a brief overview of the outcomes of the pre-2000 orderly phase of land reform (Chimhowu and Hulme 2006), focusing on beneficiaries of resettlement and on issues of productivity and income. A number of commentators agree that Zimbabwe's land resettlement programme was largely successful in targeting the poor and land hungry, particularly in the early period (Cusworth 2000; Kinsey 2004b; Moyo and Yeros 2005). From 1985 on, a stipulation was added to criteria for resettlement, that settlers should be able to use the land productively, favouring the slightly better off, for instance, those with a plough and oxen (Jacobs 1989). Provision was also made to include Master Farmers: those with agricultural experience and state certification. According to detailed, longitudinal data from the Zimbabwe Rural Household Dynamics Survey (ZRHDS), the selection process appears to have been remarkably equitable and efficient (Kinsey 2004b; Chimhowu and Hulme 2006), and there was little sign of the programme as a whole being 'coopted' by the wealthy.

The programme was also successful in general in raising marketed output, all the more remarkable since Zimbabwe suffered several droughts during the period in question. A summary of several surveys indicates that household income from cropping was higher among resettled households than in CAs (Kinsey 2004b)

Two United Kingdom government reports on land resettlement also gave highly positive evaluations of the rate of return on land reform (Cusworth 2000), an important issue for the British government as a significant funder of the programme. These studies also indicated several economic problem areas in land resettlement.

Firstly, the 1988 Overseas Development Agency (ODA, now DfID) report indicated that arrangements for credit, marketing, and input supply services were inadequate to meet the needs of settler households (Cusworth 2000: 28), causing declines in productivity and indebtedness for a number of households. Secondly, the ZRHDS data indicate that although productivity was higher in resettled than CA farms (Kinsey 1999; Kinsey 2004b; see also Muleya 2000; Palmer 2000), it did not translate into higher *per capita* incomes, mainly because over time, more prosperous resettled households attracted more members. Extra household members did not always contribute to production, although this situation did allow the economic benefits of resettlement to be spread more widely. Lastly, the resettlement programme was treated to some extent as an island with special provision and administration. Even had targets for beneficiaries been met, the poverty in CAs would likely have continued (Cliffe 2000; Stoneman 2000; Moyo and Yeros 2005). Despite some weaknesses, the pre-2000 land reform can nevertheless be seen as successful in a number of respects, including in terms of productivity.

Gender and land resettlement in Zimbabwe: the first period

How did women fare within the 'orderly phase' of resettlement? This section draws particularly upon my own research, conducted in 1983 to 1984 and the first substantial study of land reform and gender within the country. The research included individual household model (Model A) resettlement as well as production cooperatives (Model B). The study of Model A resettlement was based on 207 questionnaire interviews with wives, husbands and widows, 25 in-depth interviews, and ten key informants interviews in Hoyuyu RA (in the east) and Darwin RA (north of Harare). Additionally, twenty-five large focus groups were conducted in women's club settings within eight Model A RAs in Mashonaland East and North and in Manicaland; these contacted 650 to 700 women (see e.g. Jacobs, 1989, 1995, 2000b). I also draw on the work of Chenaux-Répond (1994) and on Goebel's Hwedza study south of Harare (1998, 1999, 2005a, 2005b).

The Zimbabwean Model A programme, like most other land reforms, allocated and allocates land to the household head. Since under customary law the husband is the head of household even in his physical absence, this process frames other gendered outcomes. Importantly, a divorced woman is likely to be asked to leave the RA as she does not hold the necessary permits for settlement, and divorce and desertion rates in the country are very high (Pankhurst and Jacobs 1988; Goebel 2005b). Zimbabwean land resettlement does, however, allow for allocation of land to female household heads, or widows.[3]

Relatively few widows or divorcees were resettled in the early stages of the programme, but by 1995, 15 per cent of all settlers were widows (Gaidzanwa 1995), a high figure in international terms. The number is high for two reasons. Firstly, a ruling (although not a law) established that customary inheritance laws did not operate within RAs, an important condition given that much feminist organising in Zimbabwe focussed on the issue of inheritance (Stewart and Tsanga 2007). Thus, a widow rather than the eldest son or the husband's lineage kin could inherit land rights or permits to stay. Since most wives are younger than their husbands, many women were able to inherit (Goebel 1999). Secondly, ROs often used their powers to intervene on behalf of women, including intervention in favour of widows' inheritance rights.

Both detrimental and beneficial effects of land reform have been noted in studies of gender relations. Women in my study cited several negative effects, beginning with an increase in workloads. The increase resulted from having more land to till, but also from increased work intensity. Relatedly, many married women and widows reported concern over lack or insufficiency of amenities within the new RAs. These included shops, access to transport, marketing opportunities, schools and clinics. It is possible that some of these problems were resolved as more amenities and infrastructure were provided, although the state partially withdrew from the early 1990s. Two other outcomes seen as detrimental by many women involved family and marriage relations. One concerned the continuous presence of husbands. The second was high rates of polygyny. The rate found in my study was 30 per cent (Jacobs 1995, 2000b) and Chenaux-Répond found rates of polygynous marriage of up to 36 per cent (1994). Although data for CAs were not reliable, it seems highly unlikely that the percentage of polygynous marriages approaches these levels. This phenomenon may indicate that many junior wives in RAs were being treated as labourers, similar to the phenomenon found by Weinrich (1975) and by Cheater (1981) in Small-Scale Commercial Farms (formerly, African Purchase Areas). Certainly, some polygamously married men I interviewed were explicit concerning their 'accumulation' of wives in this fashion and were enthusiastic about the possibilities offered within RAs. It is likely that the high rate of polygyny increased the insecurity not only of wives directly affected, but also that of other married women in RAs.

A third set of issues concerned feelings of isolation, compounded by worries about living on the ancestral lands of other people or lineage groups. Isolation was particularly a problem for widows I interviewed especially as lone women attract suspicion, not least from married women. For married women, concerns about living with unknown people had a positive side in increasing interaction with husbands. One of the studies conducted as part of the ZRHDS longitudinal research focussed on community interaction (Barr 2004), finding that resettled villagers coped with lack of ready-made kinship ties by increasing interaction and by strengthening locality-based networks. Thus, new types of communities were being formed over time and the elderly were especially active in creating new social ties.

Many facets of resettlement were more positive when viewed from a gender perspective, including aspects of property and income, family formation, and the impact of state ROs.

Turning first to property and production issues, it is evident that agricultural production and household incomes increased for many Model A households. One of the purposes of land reform is to increase rural productivity in the small farm sector. However, most case studies of agrarian and land reform report a *loss* of income for wives (see Chapter 3), largely because titles or deeds are granted to the husband as household head who often then appropriates any increased income. Additionally, wives often lose rights to their own (small) plots where this right is customary, as in most of sub-Saharan Africa. However, Zimbabwe presents a counter-example in at least two respects.

A majority of wives have been granted access to their own plots by husbands in resettlement. In my sample, 37 per cent of wives said that they had access to their own plots of land on which to grow food crops. For many, this was an improvement on the situation within CAs. Other studies have found much higher proportions of wives with access to garden plots (Chenaux-Répond 1994); for example Chimedza (1988) reported that nearly all wives had access to plots, and Goebel (1999), that 65 per cent of married women in her sample had been allocated land. Additionally, in my study one-quarter of both men and women stated that wives were allowed to keep the proceeds from sales of certain crops such as groundnuts or cotton, representing a marked improvement from their situations while in CAs. Wives' income is a highly contentious matter in Zimbabwe. Recent reinterpretation of customary law by many men means that husbands often appropriate all income, including that from formal sector work (Pankhurst and Jacobs 1988; Jacobs 1995), rather than following customary norms in this regard.

Zimbabwe was also highly unusual in that many wives' incomes increased with resettlement, contrasting with the international pattern. Wives in my sample reported high personal incomes, amounting to one-quarter of mean reported household income (Jacobs 1989). This was mediated by social class factors, with 20 per cent of women reporting having *no* income. However, wives' high average incomes indicate a high degree of intrahousehold redistribution. The finding of increased income for wives within resettlement was replicated by Goebel's study (1999), so that it is possible to write with some reliability that many wives saw an increase in their own incomes.

Female-headed households in Zimbabwean RAs were, and are, poorer in general than male headed households (Jacobs 1989), following international patterns. Chikondo, in a Msasa-Ringa RA study (cited in Kinsey 2004b), found 91 per cent of female-headed households, but 69 per cent of male-headed households, to be in the lower half of income distribution. They were more food secure, however, indicating that women on their own are often better able to retain control of uses of household income (Koopman 1997; Chant 2004). In my study, improvement in food security was a strong source of satisfaction with resettlement for both married and unmarried women, despite the problems arising from relocation.

The favourable outcomes noted relate to changes in family forms and family dynamics with resettlement. Two patterns of family formation were evident in my research: an increase in both polygnous marriage and in nuclear families. Settler families were usually two-generation families, unlike the extended three-generation lineage-based families more common in CAs. The 'familial model' of most planners was of a nuclear, two-generation household and preferably, one containing only one wife (see Chapter 3).

Despite the increase in polygamy, the majority of settlers nonetheless lived away from extended families and the majority lived in (formally) monogamous nuclear households. Seventy per cent of husbands and 60 per cent of wives in my sample said that the nuclear or 'small' family was the 'best unit in which to live'. Many people felt that extended families had caused great friction in their lives, particularly because they can entail calls upon resources. Additionally, most women and men appreciated being freer to make decisions without the input of the senior generation. For many wives, and particularly those in monogamous marriages, resettlement meant an increase in influence. In the absence of the husband's patrilineal kin and 'living among strangers', husbands may draw closer to wives and adopt a more companionate model. Later studies (Kinsey 2004a; Goebel 2005a), however, report that extended or larger families have re-formed, although not necessarily always along lineage lines. Thus, some of the effects noted here may be temporary. Married women often cited improvements in husbands' behaviour, compared with that when living in CAs. The majority of wives viewed their male partners as 'better husbands', meaning that the men drank less worked harder directed more resources to the household rather than to other women, were less violent and esteemed their wives more.

One element in this change in gender relations reflected a shift in the gendered division of labour. Even though most settlers worked harder than previously, settler men reportedly did far more fieldwork than had been the case in CAs. For a variety of reasons, agricultural tasks customarily gendered 'male' are now commonly taken over by women within CAs. In my study, husbands and wives worked together more closely in fieldwork in the RA. In particular, men assisted more with ploughing as well as with sowing, weeding and harvesting. A number of men also reported helping with 'female' tasks such as fetching water and firewood. A very small proportion of men also contributed to housework; this is notable, given the stigma attaching to male participation in domestic labour. An early village study by Henson reported similar changes (Henson 1984). Wives commonly said, 'We work together here as a unity'. Similar phraseology is reported by others. Goebel cites the phrase: 'Here it is our land, the two of us' (1999, 2005a). However, these phrases should not be taken too literally. Most women's overall work burden, involving as it does nearly all domestic labour and childcare as well as longer hours of work in agriculture, remains considerably higher than men's. In the late 1990s, only 2 per cent of wives held settlement permits (Peters and Peters 1998: 193). Nevertheless, the burden of agricultural labour is perceived as being shared more equally, and this is felt to be a highly beneficial development.

Women settlers appeared to be familiar with the image or model of a more self-contained nuclear family with a 'good' husband. This, it should be noted, does not imply female autonomy or equality. 'The history of ideas about the Good Husband would reveal continual change and yet this has not threatened the traditional authority of husband over wife' (Bell and Newby 1976: 161). Indeed, some women in my sample said that they had *lost* power or influence or else that there was 'no change' with resettlement. They offered several reasons, one being that the husband was more continually present and able to scrutinise their actions, since migration was not allowed.

Women who said that they has lost influence were most likely to be in the 'middle peasant' or self-sufficient stratum, as measured in my research.[4] They were also more likely to be in polygynous marriages. Poorer and relatively wealthier peasant women were more likely to say that they had gained influence through resettlement. Since the aim of land reform is to establish a small farming economy, the decline in influence of middle stratum women is a concern. Thus, many middle stratum women said that husbands took decisions over cropping, budgeting, and so on without consulting them at all.

The majority of women in my study nevertheless perceived that their status had improved with the *minda murefu* (long fields) of RAs. Several explanations may be relevant for this response. One is contingent: beer halls were usually much further away and opportunities to carry out extra-marital relationships, perhaps fewer. As noted, the predominance of the nuclear family may have improved wives' position. A second explanation concerns the authority of ROs. The role of ROs, who administered RAs until 2000, derived from Rhodesian community development policies. The RO's role was similar to that of the old District Commissioner and the areas do not fall under the jurisdiction of local government bodies, or District Councils. Hence, ROs had a good deal of power and were *perceived* to have power within their own RAs. In particular, they had some leeway to interpret rules applying to resettlement. Both men and women knew that their behaviour was scrutinised by ROs. Male title-holders knew that they stood to lose permits if not, as it was said, 'on good behaviour'. In practice, among the ROs I interviewed, settlers were only removed for gross misbehaviour, but the possibility of removal appeared to be felt as a deterrent. Additionally, some ROs imparted a model of 'proper' family behaviour: in some cases this consisted simply of hard work, investment in the farm, and not disrupting neighbours; for others, a notion of the wife as companion and as due respect was an element. Goebel also noted that ROs at times intervened in disputes on behalf of married women as well as widows (2005a). Even a limited view of proper behaviour tends to encourage men to be sober husbands as well as good farmers. In this instance, social and state control of a direct kind, as well as familistic ideology of a more indirect nature, may have operated to the benefit of married women. This aspect of change in gender relations is paradoxical: a 'top-down' structure operated to the benefit of some of the settlers least powerful in social terms and appeared to improve women's position. The removal of ROs in fact coincided with the onset of 'fast track' resettlement.

Overall, then, the orderly phase of land reform brought a number of benefits for married women as well as for widows, although these were not uniform.

Those women in monogamous marriages and in 'poorer' and 'wealthier' peasant strata were more likely to benefit in terms of increased influence, but most women benefited in terms of improved food security. Improvements were in the main due to state supports and to shifts in family and household relations which resulted in wives being able to exert more influence. Thus, the Zimbabwean case pre-2000 presents a broadly positive example.

The Context for 'Fast-Track' Land Reform post 2000

During the 1990s the process of land reform stalled, despite the Land Acquisition Act and despite an escalation of informal land provisioning or 'squatting'. The Zimbabwean government accused the British, substantial donors to the land reform programme, of reneging on agreements concerning financing land acquisitions. It argued that with budgetary constraints and structural adjustment, it was unable to provide the matched funding required (Lebert 2006). In the late 1990s, government provided only 15 per cent of the funding it had promised (Pallister 2000). The United Kingdom for its part alleged that monies had not been used for land reform but to hand over land to new African elites and to ZANU officials and their families (Lebert 2006: 51). In some cases, state farms had been transferred.

Although lack of funding for land reform is often blamed for the slow progress of the 1990s, other factors are also relevant. These include the priority given to commercial agriculture both by the state and by multilateral donors; corruption; and the elite's flagging commitment to resettlement (Goebel 1998). Opposition Member of Parliament Margaret Dongo[5] obtained parliamentary evidence in 2000 indicating that land leased under a tenant farming scheme from the 1990s as well as land bought from LCSFs in 1998 had gone to politically well-connected individuals who lacked farming experience and were in the main, absentee landowners (Pallister 2000). By the late 1990s, 10 per cent of agricultural land was owned by black farm owners (Moyo and Yeros 2005: 185).

Despite government ambivalence towards land reform (Kinsey 2004a), in 1998 a multilateral conference was held, including the World Bank, the UK's Department for International Development (DfID), DANIDA (the Danish International Development Agency) and NORAD (the Norwegian Agency for Development Cooperation), reaching agreement on the importance of agrarian reform for Zimbabwe and on the way forward. It is not clear why its position was not further acted upon (Hellum and Derman 2004). In any case, matters took a dramatically different turn from 2000.

The impetus for new land acquisition comes from political events in and before early 2000 rather than from a direct concern for redistribution. The events were the formation of the current opposition party, the MDC, and then the defeat of a proposed constitution in 2000, which would have greatly consolidated President Robert Mugabe's considerable powers (Jacobs 2000b; Palmer 2000). Land occupations were at least in part a distraction, as little interest in land existed during the 1990s (Hammar 2003). The political impetus for events is perhaps clearer following the 2008 elections than it was in 2000.

The official position on the beginning of land confiscations is that land invasions by ex-war veterans were spontaneous and left the government with little option but to follow their lead (Moyo and Yeros 2005). Land occupations and informal settlement have taken place since the early 1980s (Chimhowu and Hulme 2006). 'Squatters' movements indeed escalated from the 1990s with the slowdown in land resettlement. In 1998, disgruntled liberation war veterans won allocation of 20 per cent of the land designated for resettlement. However, at that time little land was available for redistribution.

Shortly after the electoral defeat of the new constitution, the ZANU government made alliance with war veterans, some of whom were occupying large commercial farms. The electoral campaign and promises of land reform became linked, since Shona-speaking[6] rural areas were a government stronghold, and Mugabe was becoming increasingly unpopular in urban areas. Much evidence points to the close involvement and orchestration of occupations by the government (Kinsey 2001; Hellum and Derman 2004). Since war veterans were in fact in charge of security including for President Mugabe's office (Moyo and Yeros 2005), it is difficult to see how the movement can be seen as independent of the state.

The period since 2000 has seen a great escalation of state-orchestrated violence, including abductions, tortures, rape and murders on a widespread scale (Amnesty International 2002a, 2002b, 2007; Human Rights Watch 2008). Violent attacks spread to MDC supporters, to anyone suspected of being a supporter, to white farmers and commercial farm workers as well as to Ndelebe-speaking people. In 2005, Operation *Murambatsvina* ('Clean up the Filth') was launched in urban areas of large cities, directed mainly at people living in informal housing who were seen as suspect. Over 700,000 homes were bulldozed and their inhabitants forcibly removed to the countryside (International Crisis Group 2005; Russell 2007). Two million people were affected, and between one-half and one million had died by 2007 from exposure, illness and starvation (Johnson 2007: 32). Approximately 3,500 houses were built to accommodate those uprooted from cities. Of these, most were given to civil servants, police, and soldiers instead (Kubatana.net 2008). Violence includes numerous sexual and other attacks on women (Hellum and Derman 2004; Goebel 2005b).

The Zimbabwean economy shrank rapidly between 2000 and 2006, by 50 per cent (Howden 2006), accompanied by hyperinflation. As noted in the Britain-Zimbabwe Society 2008 economic report (MacGarry 2008), trends were difficult to measure because of the scarcity of hard figures. By July 2008 the local Zimbabwean dollar was of little worth. A number of systems had collapsed almost entirely by mid-2007, including water provision, sanitation, traffic lights, telephone communications, and power.

The health system was operating sporadically, partly due to migration of most staff abroad as well as to power cuts and lack of medications and equipment (Ushe 2007; McGreal 2008b 18.7: 14). Rural living standards applied in Harare, with electricity nearly unavailable and drinking water drawn from shallow pits (IRIN News 2007a). In late 2008, a cholera epidemic broke out, exacerbated by the collapse of infrastructure; by mid-January 2009, over 2,000 people had died and 40,000 had contracted the disease (Reuters 2009). Much economic activity is based on barter and on the informal economy, while remittances from approximately four million Zimbabweans living abroad (Meldrum 2007) are crucial in allaying very severe poverty and famine.

It is difficult in fact to estimate the size of the current Zimbabwean population. Very large numbers of people have died as a result of famine, economic collapse and the knock-on effects of collapse of the health system (Howden 2006). Approximately 20 per cent of the adult population suffered from HIV/AIDS in 2005 (AVERT 2008a),[7] and their deaths have been hastened through malnutrition and lack of health care (Johnson 2007). Aside from those living abroad, several millions are missing from the population, with estimates ranging from a minimum of two million up to six million (Johnson 2007: 32; AVERT 2008a). The fact that the margin of uncertainty is so wide is an indication of how fragile are the country's systems. Thus, disruption, crisis and violence are by no means confined to rural areas.

Fast-track Land Reform

The confiscation of commercial farms was accompanied by a new 'fast-track' phase of land reform from 2000 on. What role has land reform played in the situation described previously? The share of land owned by mainly white commercial farmers at the time of the 2000 land invasions was 30 per cent, held in 6000 large estates. This was reduced to 3 per cent in 2004 (Goebel 2005a: 47). In 2007 the Commercial Farmers' Union represented approximately 400 farmers (Business News [SA] 2007). Late that year, Mugabe announced the end of land confiscations, although evictions have continued. Two models of resettlement were put in place. Model A1 redistributed land to small-scale farmers; Model A2 redistributed land to larger, aspirant commercial farmers. By 2004, some 200,000 people were said to be resettled (Masiiwa 2004).

The fast-track resettlement has been marred, as indicated, by much violence. Although beatings, rapes and murder of white farmers and their families have received attention, greater violence has been meted out to commercial farm workers. Suspected of siding with their employers in order to keep their jobs, they have become victims of the process rather than benefiting. Attacks on them 'were practically encouraged' (AVERT 2008a). In part, this is also because many have origins outside Zimbabwe, in Malawi, Mozambique, and elsewhere and as 'non-indigenous people' they have been marginalised in the nationalist upsurge. Commercial farms had supported two million workers, although they often worked in appalling conditions. In 2003, the number was reduced to 300,000 to 350,000 (Sachikonye 2003), and in May, 2006, approximately 200,000 people were employed on farms and plantations (MacGarry 2007).

The Zimbabwean reform stands out as one of the few in which male workers on agricultural estates have not benefited. The government's land audit or Utete Commission Report in 2003 found that only 1 per cent of the settlers had been commercial farm workers (Government of Zimbabwe 2003). MDC supporters are also excluded from land redistribution, since land is seen as a reward for political allegiance.

Little data, and no systematic data, exists concerning recipients of land reform.[8] Land redistribution has been carried out by ZANU and auxiliary organisations including militia and youth brigades; it is often acknowledged that the process has been confused and chaotic as well as violent (IRIN News 2003; IRIN News 2007b). Since 2000, discussions about numbers of government ministers, war veterans, businessmen and other influential ZANU members who have received land have continued. This issue was acknowledged by Mugabe (Masiiwa 2004), but it is possible that there will never be clarity on the recipients of land during this process (Kinsey 2004b). Although the 'landless' should receive land, there are no guidelines about what constitutes landlessness (Hellum and Derman 2004) and no comprehensive list of beneficiaries exists (Ikdahl et al. 2006). Certainly some beneficiaries have been well-off urban people. In 2004, at least 400 high-ranking people had acquired land in fast-track resettlement (Masiiwa 2004). Whole farms have been allocated to highly placed government officials (Hellum and Derman 2004: 1794) and a powerful elite has acquired much of the better quality land (Kinsey 2004b: 1663).

Despite these distortions, most beneficiaries are nevertheless poor people (Moyo and Yeros 2005), whether originally rural or displaced urban people. Scoones (2008), in one of the few available empirical studies, found that 60 per cent of a Masvingo sample were land-hungry farmers. Women as household heads have received some land in the fast-track process. The government's Utete Commission figures for 2003 indicate that 18 per cent of Model A1 (small-scale farming) and 12 per cent of Model A2 (larger commercially oriented farms) were allotted to women household heads (Hellum and Derman 2004: 1796). These figures fall short of the 40 per cent recommended in the Utete Report itself or the 30 per cent recommended by the Women's Land Lobby (Mgugu 2002).

During the period of 'fast-track' reform and exacerbating the previous economic deterioration, Zimbabwe has suffered an agricultural and food security failure. By 2008, starvation was setting in (McGreal 2008c). As the preceding discussion indicates, the state has played a role in this situation. Agricultural production in 2006 to 2007 was 30 per cent less than in 2000 and Zimbabwe's food imports doubled in 2007 (MacGarry 2008: 3). The situation is commonly attributed to the commercial farm invasions and to land redistribution *per se.* Certainly, to exit from the world (capitalist) economic system in terms of export industries even with a planned process in place would be risky. In the Zimbabwean case, no plan appears to have been created. For instance, tobacco exports were a mainstay of the economy during the 1980s and 1990s. Flue-cured tobacco was the biggest export crop, and Zimbabwe was the second largest producer in the world (Kwidini 2008). Farmers' capacity to produce the crop has been affected and some have stopped selling because of non-payment. Approximately 700,000 people were directly and indirectly supported by this industry.

Droughts have also played a part in famine, as the government stresses. However, previously Zimbabwean agriculture was able to cope with drought because it had various coping systems in place (Hamadziripi 2007). These included irrigation systems, reservoirs, and human knowledge and experience. Such systems have partially broken down with the general economic collapse, mass migration and generalised destitution. On a similar note, in 2007 elephants roamed free across the borders of Hwange National Park, often destroying crops in their way (Russell 2007).

Misconceptions exist, however, about the role of commercial agriculture in food production and food security in Zimbabwe. Current food shortages have not in the main occurred because of the collapse of food staple production of white maize in commercial farms (Sanders et al. 2007). White maize for *sadza* (maize meal) is produced mainly within the small farm sector, while commercial farmers grew some yellow maize for stock feed (IRIN News 2008b). What has taken place is that agricultural industries sustained by commercial farming have collapsed with fast-track reform. Through these, small farms had access to cheap agricultural inputs such as seeds and fertilizers. Without such inputs, white maize production has declined disastrously in the new RAs as well as in CAs.

In a slightly different scenario, wheat production has declined due to the wariness of wheat farmers about growing the crop (MacGarry 2007: 3–4), because of the government's failure to deliver inputs and to shortages of money, in turn fuelled by hyperinflation. In 2005, only 53 per cent of the target area had been planted in time to harvest. In 2007, the Ministry of Agriculture estimated that only one-tenth of the winter wheat crop had been planted (Russell 2007). Thus, farmers, including those resettled, have been affected by shortages of appropriate support, including lack of inputs as well as credit.

Much of the land acquired through commercial farm seizures has not in fact been reallocated. In 2004, the Special Affairs Minister John Nkomo admitted that government had failed to resettle 60 per cent of the land allocated to small farmers (Ferrett 2004). In July, 2006, Mugabe stated that only 40 per cent of the 11 million hectares acquired from commercial farmers had been taken over by black African farmers (MacGarry 2007: 2). Further, much of the land which has been resettled has not been fully utilised. In 2005, an experienced researcher on land reform, Bill Kinsey, was able to travel widely. His impressions give an indication of why the agricultural crisis has developed.

  i) A good deal of land viewed had been extensively burnt, resulting from increased hunting in ex-commercial farms and the use of fire to clear land, in the absence of fire breaks;
  ii) little in the way of agricultural preparation or irrigation was going on at a time in the season when this was crucial;
  iii) the location of some new farms was in areas that required preparation such as terracing; and
  iv) some farmers had planted wheat crops that grew, but they were unable to harvest these due to lack of inputs such as fuel for combines.

By September, 2005 it was too late to harvest a good-quality crop (Kinsey 2006: 1–2).

Figure 8.1 Women and children receiving maize meal food aid, Killarney, near Bulawayo, Zimbabwe, February 2008. (Photographer: Katinka Ridderbos; thanks to Internal Displacement Monitoring Centre/ IDMC)

In 2006, 68 per cent of RA and CA small farmers were not fully utilising their land (MacGarry 2007). In mid-2008, President Mugabe admitted again that only 42 per cent of land seized was 'in full use' (Rice 2008). The epidemic of HIV/AIDS plays a part as well (Izumi 2006). Many adults are ill and malnutrition hastens the onset of AIDS from HIV status.

Factors within and outside state control therefore interact. Lack of infrastructural and other support for land reform play a large part in the problems of 'fast-track' resettlement. Land redistribution without access to fertiliser, agricultural implements or machinery, credit and extension advice is not likely to succeed. The Zimbabwean situation has been fuelled by government incapacity or unwillingness to act and by general economic breakdown. Scoones *et al.* (2008) found that some A1 farmers had been able through individual efforts, however, to make productive use of land. For others, lack of knowledge of farming is an issue; people encouraged to seize land do not always have previous farming experience. Some are in any case 'new farmers' forced to relocate from urban areas by state operations such as 'Clean up the Filth'. They often found the land uncleared, and had no farming implements or cash to buy these (Russell 2007).

In a situation of food insecurity and scarcity, the government retained control of distribution of food aid by refusing independent access to NGOs or the UN (Amnesty International 2002a; IRIN News 2007b; McGreal 2008a, 2008b). This ensured that the state and state agencies should channel food aid to some but withhold it from others. Given the large numbers who required food aid, millions of people were affected.

The state attempted at once to tighten control and to reduce reliance on food aid by other 'operations' such as the unhappily named '*Taguta/ Sisuthu* [in IsiNdebele]': ('People have had their Fill'). In 2005, the Operation deployed soldiers to rural areas to oversee maize production. The army has in any case for some years been involved with food production and has been able to manipulate it (Solidarity Peace Trust 2006). Such campaigns additionally give employment to young men, as well as assisting in feeding the army (Solidarity Peace Trust 2006). Thus, agriculture has become militarised, in a situation reminiscent of the Stalinesque tactics of the Soviet Union in the 1930s. Rather than increasing production, Operation *Taguta/Sisuthu* has destroyed market gardens, as soldiers have torn up vegetables and ordered smallholders to plant maize. Particularly in Matabeleland, the scene of past atrocities, the military has undermined the basis of agricultural production and has destroyed ripe crops (Solidarity Peace Trust 2006). The process in train constitutes a new patrimonialism (Cromwell and Chintedza 2005). Patrimonialism bases belonging on blood ties and on allegiance to a ruler or despot. This has been an effective mobilising device for male citizens (Hellum and Derman 2004) but holds serious ambiguities for women.

Gender Issues in Fast-track Reform

The severity of the crisis and lack of data in Zimbabwe means that it is difficult to ascertain how women have fared in fast-track resettlement. Given that
women's rights to control property and to make decisions are far more constrained than men's, most women are likely to be in a weak economic position. Nevertheless, women's responsibilities for food provisioning remain.

The only systematic data available, as noted, indicate that women have
received redistributed land, although not in large numbers. Some women
have acquired land in the fast-track commercial model, A2. A precedent
existed, since by the 1990s, some women had succeeded in obtaining commercial farmland. They were usually well-off, educated, held the agricultural
land with husbands, and had alternative employment (Petrie et al. 2003).
It is possible that a better-off stratum of women has not only acquired but
retained land in the A2 model. However, countervailing factors threaten
property-holding for most women. In any case, their situations differ greatly
from those of poor women seeking land on which to base livelihoods.

Women have also acquired land in the fast-track smallholder programme, model A1. Some women, including widows, have acquired land
in their own rights through participating in land invasions (Izumi 2006).
In one of the sites of Izumi's study, nine of nineteen widows had obtained
land through occupations. Women have also received land as police officers, extension officers, or in other roles (Hellum and Derman 2004). The
results of fast-track land acquisition appear to be ambiguous (Izumi 2006).
Research carried out for the Norwegian Refugee Council found that women
as household heads were negatively affected in the fast-track programme
as they lacked resources for successful resettlement (Hellum and Derman
2004), particularly given absence of state support. Women are likely to be
seen as less deserving candidates for receipt of land as smallholders because
they are viewed as lacking agricultural experience and lacking assets in
order to obtain credit for inputs. Women who resettled with husbands
remained at much risk of losing their land upon divorce or widowhood.
Widows who were HIV positive were markedly more vulnerable as they
were viewed by the husbands' kin as weak, vulnerable, and not in need of
land due to their illness (Izumi 2006).

Zimbabwean women's mobilisation has centred on attempts to challenge aspects of customary law unfavourable to women through legal
means. In this regard, the campaigns' targets were similar to elsewhere
in Africa. One result of campaigning was a new law governing inheritance in 1997 (the Administration of Estates Amendment Act), which gave
women rights to inherit property from husbands and fathers. However,
several Supreme Court decisions in the late 1990s reinforced customary
law and reasserted adult women's minority status (Stewart and Tsanga
2007). Although Section 14 of the 1996 Constitution outlaws gender discrimination, a 'clawback' clause asserts the priority of customary law in

matters of adoption, marriage, divorce, burial, and devolution of property upon death (Izumi 2006: 8). This continues to be of relevance to many groups of women. From 2000 on, legal campaigns faltered in face of the government's assertion that the rule of law was detrimental to poor people (Hellum and Derman 2004). The doubts over women's legal rights to land and property if anything, however, are exacerbated in the current situation. These considerations will not disappear, and are likely to re-emerge in future, more ordered times. For the time being, issues of famine and violence as well as land access preoccupy many women. Men as well are targets of violence and women can and do perpetuate violence (Jacobs et al. 2000a; Jacobs 2008). The widely reported establishment of 'training camps' in which torture has been enacted systematically has, for instance, implicated women (Panorama 2004). More commonly, however, women have suffered through rape, torture and assaults.

Women are not always targeted primarily *qua* women; they may be displaced, attacked or murdered in other capacities such as being commercial farm workers. However, women are far more likely to be victims than perpetuators. Women of all groups are likely to be 'punished' through sexual assaults and rape. Reports of rape are very widespread, and political opposition women (in the MDC) have been particular targets (Amnesty International 2002b, 2007; Goebel 2005b; BBC 2008; McGreal 2008b). White and Asian women have also been attacked. So, too, have Shona-speaking ZANU supporters, although the latter are also most likely to be allocated land.

The violent situation that has emerged is especially threatening for women. Even where they do gain land and have inputs to farm this, they are unlikely to be able to exercise rights in such a situation. Human rights are not only economic but also involve personal security, particularly with regard to women's human rights. The new patrimonialism is especially exclusionary for rural women.

An egalitarian and transparent agrarian reform remains an urgent necessity in Zimbabwe, despite the misuse of land reform policies. Politicisation and coercion as well as lack of meaningful state support have undermined the most central purpose of land reform, to ensure food security. This is hardly a situation in which women are likely to benefit. It is also in marked contrast with the previous land resettlement programme. Pre-2000 land reform did not give married women full rights but partial reforms of customary law, improvements in productivity and shifts in gender relations pointed in a positive direction that might be built on in future. So too, does the continued resilience of some small farmers.

SOUTH AFRICA

There are many similarities in the agrarian structures of Zimbabwe and South Africa, particularly in the domination of commercial farming and the relegation of the black African population to reserves. However, the South African land reform programme contrasts markedly with Zimbabwe's.

The racialised structure of South African landholding was and is highly skewed. The 1913 Land Act and its amendments allocated only 13 per cent of the country's land to African people. After the country's first non-racial election in 1994, approximately 55,000 commercial farmers, mainly whites, owned some 102 million hectares while 1.2 million black households had access to some 17 million hectares within the former Reserves or Homelands (Marcus et al. 1996: 97). In 2000, approximately 70 per cent of the rural population had access to land. For half, access was to 1 hectare or less (May 2000: 23). As in much of Africa, the majority of the rural population is female (May 2000). Thus, rural areas are highly differentiated, with a large group of very poor black women producing mainly for household consumption and a small, mainly male elite producing on a greater scale (Lahiff and Cousins 2005).

For a variety of reasons, demand for agricultural land is not as high as in Zimbabwe. South Africa is much more highly industrialised and urbanised, with 58 per cent of the population living in cities and towns in 2005 (Globalis 2005). In rural areas, apartheid 'removal' policies created migratory populations as well as widespread homelessness and dispossession of land (Unterhalter 1987). Rural livelihoods have become diversified (Francis 2000) and for most, the primary orientation and aspiration is for waged employment (Walker 2007). Nevertheless, agriculture and land-based livelihoods remain important. One-third of rural households engage in small-scale agriculture and for nearly one-fifth, this was the prime activity (May et al. 2000). Even if agriculture is not the preferred option, the non-agricultural economy has not proved able to absorb the unemployed, and many are likely to need access to land.

The African National Congress (ANC) government elected in 1994 under Nelson Mandela instituted a land reform programme. This has three 'prongs': land restitution, tenure reform, and land redistribution. Restitution policies relate to land expropriated since 1913. The majority of claims settled have involved urban people and by 2006 had been settled in cash (Hall 2004; Cousins 2007) rather than involving land transfers. Tenure reform involves several elements, including strengthening the rights of commercial farm workers and occupants. The Land Reform (Labour Tenants) Act of 1996 and the Extension of Security of Tenure Act 1997 (ESTA) were meant to strengthen the residency rights of commercial farm workers and to protect them from eviction. However, landowners have often evicted tenants in order to circumvent the legislation. A Nkuzi Development Association survey in 2005 found that nearly one million people had been evicted from farms between 1994 and 2003, more than the estimated number of land reform beneficiaries (Hall 2007: 96). Another aspect of tenure reform concerns the rights of chiefly authorities over land allocation.

Land redistribution forms the third plank of land reform and perhaps the one giving rise to highest expectations.

The South African land reform has been market led, based on a 'willing buyer–willing seller' principle. Government does not buy land itself or select beneficiaries. Instead, potential beneficiaries must apply to the state under one of the three aspects of the programme (Hall 2004). Between 1994 and 1999, the programme took the form of pilot projects in various provinces. Low-income households were able to access a small, one-off grant or settlement land acquisition grant (SLAG), targeted at low-income households. Wives as well as husbands were listed as beneficiaries. In practice, due to the cost of land and the size of grants, grants were often pooled among households. Grants could be pooled in two types of 'legal entities', a trust or a communal property association (CPA).

From 1999 on, the direction of policy shifted to stress agricultural productivity and encouragement of an African commercial farming sector. From 2001 on, the Land Redistribution and Development (LRAD) policy was launched, altering the intent and those meant to benefit from the programme. The previous income criterion of eligibility was dropped, so that wealthier black people can apply. Grants are now to individuals rather than households, potentially benefiting women. Grants are awarded on a sliding scale with individuals expected to contribute a proportion. Although the minimum contribution was R5,000 in 2004, near-destitute rural people are unable to raise even small amounts of capital. It is still common for groups to apply for land jointly due to the high price of land, but purchases by large groups are no longer allowed (Hall 2007).

The pace of land redistribution has been slow. Between 1994 and 1999 only 1 per cent of agricultural land was redistributed (Hendricks 2003). In 1999, a target of redistribution of 30 per cent of agricultural land was set, but has been put forward to 2014. By mid-2008, 4 per cent of agricultural land outside the ex-Homelands had been redistributed and 50,000 commercial farmers held 80 per cent of land (Montsho 2008; see also Ntsebeza 2007). This total includes provision of grants to municipalities in order to purchase substantial amounts of land for communal grazing, benefiting poor livestock owners (Hall 2007). Reasons for the pace of delivery include insufficient funding (Hall 2004; Cousins 2007), although the budget for land restitution claims increased from 2004. Institutional or structural incapacity within ministries is also cited as an issue (Walker 2003, 2007; see also Thornton 2008), including in ministerial statements (Groenewald 2008).

The slow pace of delivery coupled with the Zimbabwean example led the South African Communist Party to instigate a campaign for land redistribution in 2005. An ANC policy conference in 2007 followed on, demanding more aggressive measures to redistribute land (Mail and Guardian 2007). Land invasions, both rural and peri-urban, have escalated (Sihlongonyane

2005) although are not on a large scale. In 2007, government carried out an—or, possibly its first—expropriation. This involved land which had been subject to a restitution claim and in which negotiations over the price of land had stalled (Mail and Guardian 2007). On May Day 2008, the new ANC President Jacob Zuma called for a speedup of the pace of redistribution in order to ensure food security and to fight rural poverty, not least that of rural women (Montsho 2008).

A heated debate exists over whether land reform could in fact be a motor of economic development in South African conditions if it were widened. Sender and Johnston (2004) argue that evidence for the 'inverse relation' between productivity and farm size (see Chapter 1) in South Africa is weak. Some argue that targeting of beneficiaries for land reform should include only that minority most interested in farming (Moseley, cited in Cousins 2006). A Centre for Development and Enterprise (CDE) report on land reform claimed that half of land reform projects had failed (CDE 2008: 45), an assessment confirmed by the Acting Director General of Land Affairs (van Schalkwyk 2008). Post-land transfer support is widely cited as a critical gap (Hall 2004; May et al. 2000), resulting in large-scale underutilisation of land. Women farmers' lack of decision-making power has also been cited as a factor limiting productivity, since women usually work the land (McCusker 2004). Nevertheless, land redistribution may have increased production of food crops, though this is difficult to measure. Productivity is important for both men and women, but an overly narrow focus on productivity can marginalise gender issues.

It is true that given the country's arid conditions, the loss or lack of tradition of farming for many people and the demography of rural areas, land reform would have to be part of a wider agrarian reform and development strategy in order to alleviate poverty on a large scale (Cousins 2007). Walker cautions that inflated expectations of land reform require tempering (2007: 134), not because the programme is unimportant, but in order to set this within a wider economic strategy that supports small farmers.

Gender Issues in South African Land Reform

Since 1994, South Africa has enacted highly progressive gender legislation. National legislation equalises women's and men's legal positions, protects women's employment rights, and outlaws domestic violence and sexual crimes (Goetz and Hassim 2003). A 2004 Constitutional Court decision is of relevance to land rights. The *Bhe* case ruled that the constitutional provision of gender equity takes precedence over customary inheritance law enjoining male primogeniture (Walker 2007). Implementation of laws on gender equity remains weak, however, in rural as well as urban areas (Commission for Gender Equality 2003). As elsewhere, customary law frames much of life for rural women. Additionally, the situation of women is shaped by commonly experienced risks to their health, security and social standing. For instance, HIV/AIDS prevalence is over 18 per cent among people ages 15 to 49 and 29 per cent for pregnant women (AVERT 2008b) and women are often blamed for HIV/AIDS (Turshen 1995, 2006; Izumi 2006). Rates of rape, sexual assault, and other assaults are extremely high in rural as well as urban areas (Hirschmann 1998; Magardie 1999; Moffett 2006).

In terms of stated policy, gender has been written into South African land reform policy with a degree of explicitness. The government's White Paper on Land Policy 1997 strongly endorsed the principal of gender equity and a gender unit was established in the Department of Land Affairs (DLA). Although the unit had difficulty in exercising authority (Walker 2003), its establishment did signal intent to highlight gender issues. The DLA has made three main attempts to promote women's interests, as wives or as household heads:

i) women-headed households were listed as beneficiaries on the relevant land reform documents in the first phase of the programme;
ii) in the LRAD programme a target of 30 per cent female beneficiaries was set (Walker 2005); and
iii) women are included as members of trust and CPA committees.

In CPAs, committees were meant to include a 40 per cent representation of women.

Although a DLA report in 2000 found that 47 per cent of beneficiaries were female in the 1990s, it has been heavily criticised for greatly overestimating women's participation (Bob 1999; Walker 2003: 120). Data sources concerning beneficiaries of land redistribution are in general poor (Hall 2004), though DLA figures from 2002 to 2003 indicated that the 30 per cent quota for women had been met formally. Kwa-Zulu Natal province led the way, but numbers were low at 393 (Walker 2005: 303).

Several studies in the 1990s and early twenty-first century[9] examined the gender effects of the innovation of CPAs and trusts. CPAs had to comply with the law to receive grant allocations, but actual practice does not ensure gender equity. Bob (1999), Pharoah (2001), and Walker (2003) have explored women's participation in CPA committees within Kwa Zulu Natal Province where gender norms remain highly conservative. Pharoah found that women were sometimes elected to committees based on kinship links with men, although at times as well, due to the force of their personalities. Women tend to be elected to posts such as secretaries and find it difficult to speak and to participate. Rural norms enjoin that women should not customarily speak in front of men or should do so only with modesty and deference (Cross and Friedman 1997; Mogale and Poshoko 1997). In research with 47 key informants I conducted in 2002 and 2003, a number mentioned that women were often teased or ridiculed within meetings (Jacobs 2004b), and many women were concerned about being subject to public shame.

Despite difficulties, there had been positive spin-offs from the aforementioned initiatives and from national discussions over gender equity. Policy on gender within land reform has opened up some spaces for women's engagement in negotiating for more say within households and communities. Some women felt that they had benefited from land reform, particularly in contrast with their previous lives as tenants (Walker 2005). Debates over individual versus communal forms of tenure were areas of much interest and attention (Jacobs 2004b; Walker 2005). These attempts to support gender equity have been countered, however, by other aspects of land reform. The Communal Land Rights Act (CLRA) was signed into law in 2004 but was highly contested and accords traditional leaders, considerable power to continue to exercise control over land allocation (Claassens 2003; Walker 2007). Ownership will vest in 'communities', which are vaguely defined. Although women as widows were traditionally accorded a measure of security of access to communal lands, communal tenure continues to be patriarchal. Moreover, with land scarcity, any security of access is being eroded. In the CLRA, women should comprise 25 per cent of community administrative boards, however, no penalty exists for non-compliance and chiefs are able to appoint members of their own families (Claassens 2003). Other criticisms of the CLRA are that it compromises democracy in rural areas by reinforcing the power of traditional leaders, since elsewhere in the country people are represented by elected officials. It also fails to protect members from dispossession through illegal land sales (Ntsebeza 2005; Hall 2007)

Advocates of land redistribution have also critiqued the LRAD process which favours African commercial farmers. The need to compete with wealthier people (Lahiff and Cousins 2005) and lack of state input into selection of beneficiaries means that the LRAD increases difficulty for poor women and men attempting to access land. As noted, the pace of land redistribution has in any case been hampered by a number of factors, including the need to purchase land via the market and institutional incapacity. The latter is reflected in lack of support for small farmers. If there is a future acceleration of redistribution, support post-settlement will be crucial for success.

The formal principles of gender equity in South Africa have been affirmed in land reform policy, although, these principles have not always been implemented in practice. Countervailing legislation, policy and processes may undermine stated policy. So too, does a situation in which violence against women is endemic. The partial incorporation of gender into land reform discourse has nevertheless opened up spaces for discussion, and in some cases, more concrete benefits.

DEBATES OVER CUSTOMARY LAW AND LAND TITLING IN AFRICA

A number of feminists across Africa view the way forward for gender and land rights to be in terms of individual titling. Others, as noted, warn that most women, who are poor, would lose existing rights under privatised systems (Lastarría-Cornhiel 1997). Debates concerning gender and land titling bear parallels to discussions within Latin America (see Chapter 7). However, their import is different in the context of communal land tenure systems. This section of the chapter explores the issues concerned, focusing first on World Bank initiatives then briefly examining examples from Tanzania and Mozambique.

Market-based reforms have been promoted by the World Bank and other international financial institutions since the 1970s (Toulmin and Quan 2001: 121). The expansion of export crops, often through agribusiness is a main trend globally, evident in 'new' agricultural countries such as Chile and Brazil as well as in South Africa (Rosset 2006). The World Bank's Land Policies for Growth and Poverty Reduction (World Bank 2003) sees land as a key asset for the rural and urban poor and emphasises its 'productive' use. Efficiency is seen in terms of profitability. Although the World Bank's past policy emphasised establishing freehold tenure, it has revised this with regard to African contexts. It recognises that in Africa, individualisation of land title has led to increased land concentration and landlessness and that African land systems are not inflexible (Bruce and Migot-Adhola 1994). Despite this partial backtracking, the market model still serves as an ideal type (Fortin 2005).

A number of commentators argue that individualisation and privatisation of land titles is likely to undermine women's position, especially that of wives. Where land becomes commodified, it gains value and is less likely to be used for food crops. As has taken place widely with cash cropping, women have lost control of land as it gains in value.

Using land as an asset meant that it can be used for collateral and therefore can be alienated. "The World Bank's report is curiously silent about the consequences for households defaulting on loans, having used their land as collateral." (Manji 2003b: 105). In many instances, including in the West, wives are unaware of husbands' actions with regard to property transactions, even in the absence of customary law. Thus, women whose husbands have used land as collateral for other enterprises might find themselves farming mortgaged land.

Registering or titling land is likely to result in loss of informal rights that women may hold, as has taken place under privatisation schemes (ActionAid 2005). Privatisation has broken down support mechanisms such as cooperative work and dedicated communal resources that assist poor households in times of need. Land registration is likely to weaken local institutions providing some security to community members (Hilhorst 2001). The 'African land ethic' does traditionally include concern for subsistence needs of all community members (Cross 1992) and few of the poor can retain land in an open market (Marcus et al. 1996). Thus, elites, mainly male, are usually strengthened through privatisation. Some women, however, will be able to acquire land, especially if they are educated or middle class or have access to capital (Petrie et al. 2003; Razavi 2006). However, this will not apply to most South African rural women, who remain impoverished.

A number of women's groups have nevertheless called for individualisation of land titles for women in order to secure land access (Mwagiru 1998). Contemporary circumstances have meant that access is increasingly insecure under customary law. For Nigeria, Abdullah and Hamza write, 'Ownership without independent rights is no ownership at all' (2003: 172).

Thus, titling might go some way towards giving women a measure of economic power and autonomy (Agarwal 2003) lacking in customary law, especially if they were able to have individual rather than joint titles. Women's rights are often linked with land titling. However, as the aforementioned critique of credit and land markets indicates, such a move would also 'free' women to compete in the marketplace. Many would be at risk of losing land if they managed to acquire it. Although this is similar to the position of poor men, women face a raft of additional constraints such as lack of access to credit and inputs; lack of command over labour, childcare, and care for family members; and backlashes against them. Conversely, as noted, customary law is highly discriminatory. Widows and divorcees in particular are often left impoverished and married women's rights are highly contingent. With weakening of communities, they become even more dependent upon their husband's goodwill. Thus, a conundrum exists with little consensus in this regard among land advocates.

Two country examples from Tanzania and Mozambique give an indication of the arguments on both sides. In both cases, there have been attempts to improve women's land rights. Tanzania has emphasised the market and individualised title whereas Mozambique has opted for communal rights. These cases indicate the complexity of the issues involved as well as the political splits that can take place.

Tanzania

In Tanzania, the 1999 Land Acts (in force as of 2001) were the result of complex negotiations between the state, land lobbies, and women's groups, constituting a delicate balancing act between statutory and customary law (Ikdahl et al. 2006: xi). The acts strengthen local customary institutions but seek to erode the influence of discriminatory practices and to strengthen the position of women. For instance, the Village Land Act establishes quotas for women's participation on village land councils. Additionally, spousal consent is needed for disposition of land. A (separate) 1990 court judgment found that a woman was entitled to sell land she had inherited (Ikdahl et al. 2006: 40).

Serious political splits occurred in Tanzania over land tenure issues, an account of which is given in Tsikata (2003). A land coalition formed in the late 1990s to ensure a progressive outcome in negotiations over land, but soon had to face serious differences between coalition partners (Tsikata 2003). These occurred over who should control titles in land; how discriminatory customary law should be reformed; and the amount of power to be given to the state and to village land management committees. Tensions also occurred over donor funding. The issues outlined here were mirrored in increasingly rancorous differences, ones which are relevant in other national contexts.

The Presidential Commission on Land Matters, under the chairmanship of Issa Shivji, was set up to review land legislation. Instead, its recommendations supported 'evolutionary' change to customary law. Some felt that it, as well as feminist anthropologists, overstated the strength of women's land claims within customary law. There are echoes here of the problems that peasant studies have in considering gender issues (see Chapter 2). The Gender Land Task Force (GLTF) focussed on joint ownership and registration of spouses' titles rather than the concerns of land titling that preoccupied the Commission. The Ministry of Community Development and Women's Affairs commissioned a national study which found that rural women were enthusiastic about individual land titling, thinking that it would lead to greater security (Tsikata 2003).

Following the Land Acts, the Tanzania Women Lawyers Association (TWLA) held that a number of gains had been made, including the possibility of registration in both spouses' names; equal representation of women on relevant village committees; the right for women to register an 'interest' in any land transactions (Manji 2003a); the right to acquire and register land in her own name; and use of gender-neutral language in the Land Acts (Tsikata 2003: 173). Other feminists criticised provisions on customary law as unduly weak. Although discrimination was technically outlawed, the onus is placed upon individuals to challenge such discrimination, and such action by a wife is likely to be seen as hostile.

The Shivji/Lands Commission alleged that women's groups were used by the state to divide civil society ranks since their demands did not threaten the free market paradigm (Tsikata 2003: 174). It should be noted that accusations of 'splitting' or division are charges against feminists that have been employed historically. The Shivji Commission argued that equality with men was necessary but would not create equitable land access, since both sexes faced the prospect of landlessness with titling (Tsikata 2003: 176). Women lawyers leading the GLTF replied that 'evolution' would take too long. In any case, there appeared to be no guarantee that moves to alter unequal gender power would take place.

Following the Land Acts, implementation of gender equitable provisions has been slow. For instance, husbands often sell homes and land without the wife's consent (Ikdahl et al. 2006: 39) and women's rights are often made contingent on the husband's agreement (Odgaard 2002: 82). The main aim of legislation, however, was to strengthen the land market. Perhaps predictably, gender concerns have been sidelined for this aim. Although NGOs attempt to spread information about women's rights, most women remain unaware of these, and most women cannot afford the costs of registering (Odgaard 2002). Gender initiatives have declined in visibility following the high-profile debates (Ikdahl et al. 2006). Nonetheless, the legislation can at least serve as an example of good practice in statutory terms.

Mozambique

Mozambique has arguably the most smallholder- or peasant-friendly land law in Africa (Palmer, 2005). The continued effects of displacement, sexual abuse during wartime, and other effects of conflict for women have been serious. It was hoped that improved land rights would build some stability into women's lives, post-conflict (Waterhouse 1998).

Following the end of the long and violent civil war in 1992, external agencies such as the United States Agency for International Development (USAID) advocated privatization of land as had occurred elsewhere. It is relevant that in Mozambique no serious land shortage exists and accordingly, no urgent calls for land reform (Hanlon 2004). Although Mozambique had a history of directive top-down governance, a wide consultation concerning land took place. It was remarkably open, and was followed by a radical campaign to raise public awareness. The *Campanha Terra*, mobilising around 15,000 volunteers, worked imaginatively, using many types of media. The resulting 1997 Land Law affirms that land is owned by the state and cannot be alienated. It gives communities as well as individuals rights over land on the basis of occupancy. The communities have the right to delimit and to register their lands, including fallow and reserve land (Hanlon 2004; Palmer 2005).

Debates on land and on gender and family law have been carried out rather separately, as is common. Urgent land debates concern how to deal with emerging land markets, land-grabbing by the wealthy, and the need to raise productivity (Hanlon 2004).

However, the Land Law does attend to gender issues. It affirms that land use titles can be given to men and women and attempts to give women more decision-making power in land matters (ActionAid, 2005; Gawaya 2008). Another piece of legislation, the Family Law of 2004 also has great relevance for women's land rights, stipulating that women can inherit land. Rights in land after divorce are also addressed. Other key issues relate to the codification of different forms of marriage and the principles of equity between the sexes in property as in other matters (Ikdahl et al. 2006: xii).

Due to a number of factors, women have strong use-rights in Mozambique, perhaps especially in the north. Customary practices vary widely. Both patrilineal and matrilineal patterns exist in the north, whereas Islam is practised more commonly in the south. However, a study in Nampula Province (Bonate 2003) found that customary law was of far more importance than religious or other factors in determining land rights (see also Abdullah and Hamza 2003), particularly as the state had little control over rural populations during the civil war.

As elsewhere, however, women's informal rights have their limits. One woman commented to a researcher:

> When women are working, the men approve that the women have their own pieces of land, but when it comes to harvest time, the men say: 'You are my own, and so whatever you are harvesting is also my own' (Norfolk cited in ActionAid 2005: 58).

A concern with the Land Law, then, is that the definition of 'community' supports normally male traditional authority (Hanlon 2004).

A comparative African study estimates that Mozambican law comes closest to meeting stipulations of a human rights-based approach in granting women property rights, and in the context of collective property rights (Ikhdahl et al. 2006). Nevertheless, there exists a lack of knowledge about how the multiple and hybrid contemporary laws in Mozambique have impacted locally upon rural women. Many women are still unaware of their rights, even after a concerted information campaign (ActionAid 2005). This is due in part to low educational levels, the continued impact of conflict, and to widespread poverty.

This section has emphasized the efforts made to incorporate woman-friendly stipulations concerning property, land titles, and land rights in two cases. However, it is clear that these attempts have met with a range of constraints. Many rural women are unable to utilise statutory law where favourable provisions do exist. As is now well-documented, this is due to a lack of knowledge about new laws; lack of resources to attend court; and lack of education and confidence, meaning that it is difficult to present a case (Kazembe 1986; Odgaard 2002). Moreover, many lack confidence in male-dominated legal settings, local or otherwise (Banda 2005). The weakness of many national states is also of importance. Even in situations in which a case is brought to court and a favourable judgment reached, the law still requires implementation, implying a strong institutional infrastructure and accountability (Banda 2005; see also Chapter 9). Thus, even where gender-friendly legislation such as that in Mozambique exists, it may remain 'on the books' (Gawaya 2008).

CONCLUSION

The examples in this chapter indicate a number of factors of importance in the analysis of gender within contemporary African land reforms. The first is the continued importance of customary law and practice, which frames the lives of most rural women. The brief discussion of debates over land titling and gender indicates the dilemmas involved in attempting to change traditional arrangements. The risks within market-oriented strategies include economic marginalisation and loss of land. Those within customary law include discrimination and dependent status in terms of access to and rights over land for both wives and widows.

Current debates over land reform focus on state-backed redistribution versus market-based reforms (see Chapter 1). The South African example indicates the limitations of market-based land reforms as well as issues entailed in compromise with chiefly authorities and customary law. Although the latter in general marginalises women, it may also provide a safety net for the poorest.

The Zimbabwean example is sometimes taken as invalidating all claims to land redistribution. The situation at time of writing is indeed one of economic and social meltdown but this has been only partly a result of the mishandling of land reform, which has attended neither to equity for smallholders nor to agricultural production. The impact of wider violence has been particularly negative for women, whether it emanates from the state or ruling party or is more normalised criminal or domestic violence. Nevertheless, the land reform pre-2000 had many positive features and was one in which women made a number of gains, not least because of the suspension of customary law in certain respects.

Lastly, these examples indicate that positive outcomes for women within land reforms require democracy, transparency and commitment to gender equity. This in turn implies a democratic state capable of implementing policies (see Chapter 9) and which takes women's claims seriously. Land reforms can be misused and misapplied with tragic consequences, but they continue to offer the hope of new possibilities and prospects.

9   Conclusion

The examples of gender and agrarian reform discussed in these pages present a conundrum, a set of dilemmas for policy and practice that are not easily resolved. This concluding chapter revisits the different trajectories of agrarian and land reforms. It then discusses conditions under which more equitable land reforms might take place and posits potential ways forward.

THREE TRAJECTORIES OF GENDER AND AGRARIAN REFORM

Agrarian and land reforms have either redistributed land to individual households or else to collectives. 'Agrarian' and 'land' reforms are sometimes distinguished, in that agrarian reforms have a broader scope of social transformation; in practice, the terms often overlap. What is usually meant by agrarian or land reform is redistribution to households in order to form a viable smallholding. This is what most agrarian movements have called for, and land reforms promise greater rural democracy as well as improved productivity and therefore, increased food security (see Chapter 1; Jacobs 2003). As outlined here, evidence indicates that women have experienced systematic discrimination within land reforms. Moreover, this has taken place across world regions (Chapter 3).

Despite the widespread nature of this phenomenon, regional and cultural factors play a part in the emergence of differing configurations of gender relations. For instance, women have been marginalised in many Latin American contexts of agriculture until recently because it has been seen as a male activity. Domestic ideologies also play a part (see Chapter 7). Beliefs about women's lack of suitability for agricultural work or their inferior nature are changing but persist within many rural areas. In sub-Saharan Africa, in contrast, women are the main agriculturalists. While women's large labour inputs do not *per se* translate into social power, it does mean that they have had a relatively stronger position which in some cases has fed into improvements within land reforms (see Chapter 8 on Zimbabwe). Nevertheless, cross-cultural analysis indicates that women, and particularly wives, lose a degree of status and control with land reforms, despite other positive outcomes (Chapter 3).

Campaigns for redistributionist land reform have rarely taken up gender equity. I argue that this is not accidental but is in part because of difficulty in dealing with the issue. Pointing out gender subordination within the household model threatens the idealised picture of peasant households as harmonious and unified entities (Chapter 2; Folbre 1984). This is evident even in recent movements committed to equity such as the Brazilian Landless Movement (MST) (Chapter 7). Apart from those launched by state socialist societies and those in East Asia (e.g. Korea and Taiwan), relatively few large-scale land reforms have been enacted globally. Struggles for agrarian and land reforms have often been bloody (Borras 2005; MST 2004, 2005; see Chapter 8 on southern Africa) and have frequently ended with partial and fragmentary concessions. Thus, there may be a tactical tendency to avoid 'problematic' issues that threaten to split social movements. Even in late 2004, the first *Foro Mundial sobre la Reforma Agraria* (FMRA; World Forum for Agrarian Reform) in Valencia, Spain, allowed little space for discussion of gender issues.

One type of agrarian reform established collective units, either state farms or production cooperatives (Chapter 4). In fact, the greater part of land redistribution has taken place either through collectivist reforms or else the redistributions that have followed (Chapters 4 to 6). As outlined here, these were favoured by state socialist governments for a variety of reasons. Collective farms dealt with the political issue of the peasantry, seen as individualistic and potentially petty capitalist. Collectives also retained state direction and control over food production, considered necessary to feed the cities. Gender equity was envisioned as an outcome of collectivisation, but it was assumed that this would simply follow from the removal of women's social labour from the home and their payment in work points. As stressed, women's domestic or reproductive labour was (with the exception of the Great Leap Forward in China and the 'nurturing law' in Nicaragua) viewed as outside legitimate policy concerns. Agriculture was assumed to be most efficient if organised in large industrial-like units. As several chapters in this book attest, this was an error of great proportions, and one that was repeated even after the initial Soviet experience (see Chapters 1 and 4). For a variety of reasons related to incentives to work, the need to invest in farming, the inefficiency of top-down approaches, and the detailed and complex care and attention needed for many crops, collective agriculture rarely proved capable of increasing production levels. Although not disastrous in all cases, it was in at least two. The USSR and China (Chapters 4 and 5) suffered large-scale famines as a result of the coercive manner of collectivisation or overly rapid reorganisation. *One* of the factors of relevance here was gender. Many male peasants resisted collectivisation—usually passively—in part because of the loss of control over the labour of women in their households. As Stacey put it (1983), the vision was one of a small-scale 'peasant patriarchy' rather than a large collective in which women had rights to earn work points, albeit on an unequal basis. Similarly, in Viet Nam and Hungary (see Chapters 6 and 4, respectively), collectives were unpopular among men but more popular among women including female household heads.

At its starkest, this scenario counter-poises greater gender equity within collective forms against greater food security and more productive agriculture but with strengthened gender subordination. This is, of course, oversimplified. Land reforms based on a household model may fail as well (Chapters 1 and 8), often due to lack of sufficient state support (Christodoulou 1990; El-Ghonemy 1990; Barraclough 1991; Thiesenhusen 1996).[1] Nonetheless, the issues of food security and the importance of women's labour within land reforms are real. Perhaps it is not surprising that collectives have lost out, given male power in rural society, the need for food security for both women and men, and emotional attachments to traditional ways of life and social organisation—even where such ways of life exist mainly in imaginaries.

In the last two decades, other policy options have been put forward. With neo-liberal economic strategies and the turn to the market, the possibility of state-backed redistributionist reform has been downplayed. The neo-liberal turn has impacted on discourse, so that the meaning of reform itself has been altered, harnessed in aid of further privatisation and of land titling. Many agrarian social movements (see Chapter 1) have rejected this use and have reaffirmed the importance of state-backed land redistribution (Borras et al. 2008). However, these have had little effect on moves to title land.

In contexts in which land rights have traditionally been collective—e.g. sub-Saharan Africa and indigenous areas of Latin America—such moves have been particularly emotive. Neo-liberal policy promises that titling will be 'gender friendly' in that women as well as men can receive titles. Indeed, as the work of Deere and León (2001) demonstrates, in Latin America women have made gains either directly through titling or else less directly through inheritance. However, this is within a context in which most land is already privatised. Discussions of titling do not always address the question of what is likely to happen when the poor use land as collateral to raise loans, as they are encouraged to do (see Chapters 1 and 8; Manji 2006). Women often lack the power to obtain credit or to command labour, and they are often less able to use social connections than are men. These circumstances mean that they feature disproportionately among the poor. Thus, they are even more likely than men to lose any land obtained.

Where collective rights already exist, movements defending collective land rights may end up being counter-poised against those arguing for individual rights on the basis that this will promote greater gender equity. For instance, it is sometimes argued that women have been so discriminated against and overlooked within customary systems that the system has to be dismantled in order for women, especially wives, to attain rights. Many African women have mounted new movements to eradicate customary land tenure practices and to fight for individual rights and therefore greater control over resources (Tripp 2004: 2). Mutual misunderstandings sometimes result. In the Tanzanian example (Chapter 8), feminists did not (or did not immediately) take on board the wider issues facing the community or smallholding agriculture. For its part, the Lands Commission which supported the continuation of customary communal tenure found it impossible to address gender issues or to treat these as serious concerns (Tsikata 2003).

Arguments for privatisation have thus been linked discursively with gender and discourses of liberation while those for communal systems have been intertwined with tradition and (sometimes) with implicit male privilege. That the arguments have been framed in this way attests not only to discrimination within traditional land tenure systems but also to failures of agrarian movements, and of progressive movements more generally, to address gender divisions. This omission has a long history in literature on land and agrarian reforms. It reflects the weakness of movements for agrarian reform. Such movements have received little attention (apart from the negative attention afforded to Zimbabwe) and their main actors, smallholding peasants, the landless, and land hungry, hold little social power. As noted, where strong movements occur, they are often bloody and results in terms of land distribution and productivity may be uncertain. Thus, the pressure not to complicate issues or to weaken the movement may be great. The tendency has been to address women's public roles and participation, but to ignore the crucial but contentious role of women's subordination within households. Given the general context faced by agrarian reform movements, the omission of gender issues is in part comprehensible. However, it is important to disrupt the normalised, implicit understanding that women in the movement should not raise feminist issues. Indeed, the view that gender is a distraction from the main matter of the class struggle has been evident in the MST in Brazil (see Chapter 7).

POSITIVE OUTCOMES AND THE STATE

Rural women's social positions are to some extent amenable to change through policy interventions, despite underlying structural factors within peasant households (Chapter 2). Some measures within household models of land reform have improved women's positions and given 'space' for more negotiation. As noted, greater food security is a main aim; where this occurs, it improves women's as well as men's lives (Chapters 1 and 3). Many women approve of the nuclear family arrangements usually promoted, especially if the previous systems were patrilineage based. Smaller nuclear families are seen as increasing women's influence within the home (Chapters 3, 5, 6, and 8), despite the risk of 'housewification'.

It is easier for social policy to target women *qua* household heads than as wives. Even where they are impoverished, as is common, they can

more easily act and take decisions. Most agrarian and land reform programmes today do include female household heads as target groups. The
danger is that female household heads are made to 'stand in' for gender
as a whole (Jackson 1996; Chant 2004), signalling a lack of attention
to gender relations within marriage or to the situations of non-married
women who do not head households. Nevertheless, some efforts have
been made to include wives not only as 'housewives' but as agriculturalists in land reform programmes. In the past, the Zimbabwean governments's agricultural extension service (AGRITEX) gave extension advice
to wives as well as husbands and made efforts to recruit female staff.
The South African land reform programme has insisted that women form
at least one-third of committees governing communal property associations and trusts (Chapter 8) and despite resistance, there is some evidence
that the presence of women on committees has an impact. Discussions
about extending credit to women as well as to male farmers (Gawaya
2008) show a will to include women. In transitional contexts in Viet
Nam (Chapter 6) and China (Chapter 5) wives' rights to land allocation
strengthens their position.

Women's (and men's) economic options have changed in some instances
because of the development of possibilities outside agriculture. As stressed
in the Introduction, the norm for most rural people is now to diversify
livelihoods, but this is usually an indication of poverty as well as of entrepreneurship. An exception is China (see Chapter 5) with its encouragement
of decentralised and courtyard industry. This has permitted a number of
women to combine agriculture with paid industrial activity but has not
been replicated outside that society. The UAIMs in Mexico (Chapter 7)
established agro-industrial complexes to benefit women; they were widespread but halted after the early 1990s counter-reforms. Nevertheless, they
may provide a useful model. Alternatives such as migration to export processing zones entail leaving the agricultural sector.

It is notable from the preceding examples, as well as other discussions in
this book, that where improvements in women's position within individual
household land reforms have taken place, these have mainly been from concerted government action. Actions have taken the forms of legislation; of
policy specifically targeting women; and of establishing alternatives outside
agriculture so that women are less encapsulated within the household. In
order for gender equity in land reforms even to be placed on the agenda,
it appears to take a state committed to redistribution and to alleviation of
poverty. It also necessitates a political culture in which measures for gender
equity are highlighted. This combination does not occur often, particularly
given backlashes against feminist ideas (Faludi 1992)

Land and agrarian reforms are often thought of as policies pertaining
to past eras. If this were the case, problems of marginalisation of smallholder women's interests could be sidestepped because the overall necessity
of policy would have receded. However, current food shortages, escalation

of food prices, concerns over environmental threats and unsustainable agriculture, over the conditions of work on plantation agriculture, and the encroachment on peasant land for cultivation of crops such as soybeans, mean that the agrarian question will not conveniently disappear. Nor will the question of gender, which remains central despite the discomfort it elicits. If some still wish to 'remember' an agrarian past without 'gender troubles' (to appropriate Butler's [1990] phrase in another context), the reality in the twenty-first century presents a different context, unlikely to lend itself to burying questions of gender subordination and women's rights.

The principle put forward in this book is that women must be accorded the same rights, including land and property rights, as men in their societies. Where land rights are privatised, women should be able to hold land individually. Where rights are communal and males are favoured, gender equity must be written in even where this is against tradition. With this in mind, the last sections turn to brief consideration of some possible ways forward. There are no easy solutions to the dilemmas posed in these chapters. However, some authors have considered the policy implications of approaches based on concepts of human rights and of use of international instruments. I also suggest that labour based organising may provide useful avenues.

WAYS FORWARD?

A proposed strategy for uniting gender and communal rights is the human rights based approach (HRBA). Formulated in the context of debates over communal and individual land rights in east Africa, it might have wider applicability. Ikdahl et al. (2006) posit that this approach can be an effective strategy in securing women's land rights while also preserving group rights. The principles of the HRBA include:

- recognising links between sustainable human development and human rights;
- seeing the individual as a central actor;
- focusing on *rights* rather than needs; and
- setting out legally binding frameworks of individual and group rights with obligations for national governments and international bodies.

The development of these principles would entail non-discriminatory access to land and protection of land rights; the need for gender equitable land reforms in public and private spheres; equal participation of women in formalisation processes as well as within land reforms; monitoring; and accountability (Ikdahl et al. 2006: xi–x).

The HRBA argues that entitlements to livelihoods and resources would place a brake on neo-liberal strategies that might deprive communities of resources, as the right to access to resources is part of human rights.

Another strategy also makes reference to international human rights. The International Covenant on Economic, Social and Cultural Rights (ICESCR, in place since 1976) has much potential for curtailing current encroachments upon economic rights (Elson and Gideon 2004). Feminists have focussed on the ICESCR as an arena for activism, both in terms of improving the ICESCR as a normative framework and in using it to name and shame errant governments. The Association for Women in Development has pointed to the advantages of a discourse around economic and social rights. The ICESCR is of interest to feminists for several reasons (Day and Brodsky cited in Elson and Gideon 2004). Firstly, it addresses practical material conditions and as such is of great interest to most women—including poorer rural women. Secondly, for women a division between rights to economic security and to personal liberty is purely artificial (Jacobs 2004b) as is clear for instance, in the case of violent abuse. Thirdly, the ICESCR precludes equalisation downward (as in the current 'race to the bottom' often cited as a consequence of deregulation). However, in a world in which economic power is increasingly concentrated, this would be difficult to enforce. Nevertheless, the strength of the ICESCR rests in deploying a discourse of rights, including *economic* rights. Relatedly, the United Nations (UN) has recently appointed a Special Rapporteur on the Right to Adequate Livelihood (Pena et al. 2008).

Although legally based campaigns have great strengths, they can have an unduly individualistic orientation. Nevertheless, as Razavi (2006) argues, changes in statutory law are worthwhile, not necessarily because they always result in change on the ground but because they provide an aim. In this respect as in others, there are few substitutes for social mobilisation as a way to secure rights. Ideally, gender rights would be viewed as central by agrarian and land reform campaigns, particularly given widespread feminisation of agriculture. There is some indication of movement in this direction; for instance, a recent collection concerned with mobilisations against neo-liberal reforms includes a chapter on gender commenting on contradictions between indigenous and gender rights (Montsalve 2006). The International Coordinating Committee of *Vía Campesina* has equal membership of men and women (Borras 2008). However, it remains the case that agrarian movements have often found it difficult to grapple with the complexities and disruptive nature of women's demands for equity in land reforms and land rights.

If this remains largely the case, is mobilisation via labour movements another potential avenue? Certainly, in the cases of Nicaraguan and Brazilian land reforms, labour movement-based organisations with strong feminist involvement have raised gender issues more strongly and effectively than have agrarian movements (Chapter 7). In South Africa, the strength of trade unions and of a feminist presence in some of them has had an indirect impact on agrarian policy and meant that gender is a concern within land reforms.

It may also be useful to consider the International Labour Organisation's (ILO) policy of 'decent work' with reference to gender and land rights. The agenda states the primary goal of the ILO today is to 'promote opportunities for women and men to obtain decent and productive work, in conditions of freedom, equity, security and human dignity' (ILO 1999). The 'decent work' framework is stated to be central to poverty reduction and should include equality of opportunity and treatment for women and men. However, this framework is oriented mainly to formally organised workplaces or at least ones which pay wages to workers. It is worth considering whether it could be broadened to include other sectors including informal sector traders, agricultural workers, and small-scale farmers (Cameron et al. 2006). Because redistributive land reform takes place through state-regulated programmes, the ILO approach might offer an opportunity to regulate the work of wives on farms, as well as other less-advantaged household members.

The brief discussions above can only indicate some potential ways forward in dealing with the complexity posed by analyses of gender and agrarian reform. The implication is that it is necessary to use national and international instruments or strategies mobilising labour movements to avoid repeating mistakes of the past.

Despite the failures of collectivisation, one of its great advantages was that women were treated as workers, and both men and women received either a wage or an allocation of the crop and proceeds. This is unlikely to take place within smallholder farming. However, is there a way within state-run land reform areas to gain the advantages of greater incentives in agriculture and closer relations to farming that individual household farming brings, while at the same time sidestepping its gender implications?

CONCLUDING THOUGHTS

In conclusion, three further observations are in order. State-backed agrarian and land reforms do present opportunities for formulation of new types of relations and regulatory regimes, but these must go beyond simply passing legislation.[2] Firstly, any rights that women gain within agrarian reforms, such as co-titling or joint listing on land permits, must be enforceable. As Cousins (1997) puts it, rights must be 'made real'. There are many examples of beneficial gender changes that remain on the books. Reforms in favour of peasant women are particularly difficult to enforce because of the geographical spread of households. Elson stresses the importance of accountability: to have an entitlement implies access to a transparent process. If a person's access to a resource is at the arbitrary discretion of a public official or is dependent upon patronage or the goodwill of a husband, then access to the resource is not of right (Elson 2002). Entitlements, then, imply accountability. Many legislative changes founder on a lack of accountability and are never effectively enforced, particularly in the case of gender equity.

Secondly, some type of representation for rural women is needed in which women's needs and demands can be prioritised. In state socialist societies, usually some kind of state-led representation exists, such as the Women's Union in Viet Nam or the All-China Women's Federation, as well as representation on collective bodies (Chapter 4). They have not adequately substituted for lack of civil society movements but did provide some voice and scope for redress. Women's movements now flourish in many parts of the world (Eschle 2001; Therborn 2004; Jacobs 2004a; Harcourt 2004; Ferree and Tripp 2006) but are usually urban in membership. Relatively few movements take up land and land reform as important issues. (However, see National Federation of Peasant Women [AMIHAN 2004] in the Philippines [Lindio-McGovern 1997] and the remarkable Bodghaya movement in Bihar [see Manimala 1983; Kelkar and Gala 1990; Jacobs 1996]).

Peasant smallholder women have no organic or ready-made means of organising. Aside from practical issues such as transport and lack of time to meet, their homes and workplaces are the same, and the director of work may be the husband. Where rural women's movements do exist, they often require practical and organisational support (Gawaya 2008). Where they do not, urban women's movements or rural trade unions might help facilitate activism. Some feminist organisations with a labour movement orientation have understood the need for linked strategies in thinking of women's needs and of workplace rights (Hale and Wills 2007). Creative solutions must be sought.

Thirdly, effective land rights also need to be backed up by other legal measures. In particular, women seeking to assert land and property rights often face violence from husbands and their relatives or sometimes from their own (natal) relatives. Violence is the most immediate way of divesting women of any new property rights and of intimidating them so that, even where they keep land, they are not able to exercise its use (Jacobs 1997). Thus, ensuring women's economic rights may entail the enactment and enforcement of laws in non-economic realms in order to take account of the complex circumstances many women face. A uniting of concerns for bodily integrity and protection from violence with concern for economic rights is needed.

Land redistribution programmes remain of importance for many rural as well as urban people, and women figure large among these. While state-backed programmes have often marginalised women, particularly as wives, they also present opportunities for change. International movements and instruments provide opportunities for use of international human rights based instruments, although it would be naïve to assume that these will not be contested.

Agrarian and land reforms contain potential to increase rural democracy. For this to be fulfilled, the errors of the past must be recognised.

Many land reforms have affected women's lives and positionings detrimentally. Gender issues have proved too complex and contradictory to be encompassed within populist movements which focus only on the peasantry or for movements acknowledging only social class divisions. Feminist analysis and feminist informed policy is not an optional add-on with regard to land issues. Gender relations are affected centrally by agrarian reform policies, and gendered agency in turn affects the direction of agricultural and state policies. Land reforms are not democratic unless women achieve rights, autonomy, and better life chances within them; they as well as men are 'tillers of the earth'.

Notes

NOTES TO THE INTRODUCTION

1. I studied Zimbabwean land reform and resettlement in the early 1980s (see Jacobs 1983, 1989). This was the earliest study of the subject on gender within Africa and also provides primary data for a later chapter. However, most of the book draws upon secondary sources.
2. For instance, there is virtually no literature available in Korean or in English on gender effects of agrarian reform in South Korea (Song 2009), although the agrarian reform there was one of the most successful in the world (Sobhan 1993; Griffin et al. 2002). The Philippines has highly inequitable landholdings, an agrarian reform movement, and rural women's movements (see Chapter 9). However, any material available on women in the (limited) land reform there is likely to be in unpublished reports.

NOTES TO CHAPTER 1

1. Fewer societies were 'feudal' than is sometimes assumed. Analyses influenced by Stalin held to the necessity of a unilinear path of 'development' from feudalism to capitalism and then socialism. Thus, the term 'feudalism' is applied to a whole range of societies whose structure is different to that of Europe and Japan—e.g. India (Anderson 1974).
2. Griffin, Kahn, and Ickowitz have also been criticised for ignoring the issue of class differentiation. For instance, where wealthy peasants are able to raise credit, they can use or even monopolise new technologies. In this way they can achieve greater productivity on larger plots (Dyer 2004).

NOTES TO CHAPTER 2

1. 'Horticulturalist' is the term usually assigned to agricultural producers in simple, non-state societies. Cultivation is most usually with hoes rather than with ploughs, and often shifting cultivation or swidden agriculture is practiced. Land is often communally held or owned, although used by the group or individual household. The wider social division of labour is (or was) based on kinship, usually, on a lineage basis. In the past, markets were weakly developed and cultures were seen as largely homogeneous (Fallers 1961; Post 1972). This type of society is also called 'tribal'; this term has fallen out of fashion due to use in other contexts to refer to ethnic affiliations.

2. Chayanov's work contains a theory of 'demographic differentiation'. This is a cyclical socio-economic differentiation based mainly on demographic factors and particularly on the number of able-bodied adult men and women in the household. It posits that differences in wealth among peasant households depend on demographic differentiation rather than class differentiation.

3. For Shanin, the global similarity of peasantries includes the following features: the peasant household; the existence of villages, local markets, and interaction with lower rungs of state authority; and specific reactions to processes of structural change such as collectivisation and commercialisation (Shanin 1990: 54).

4. Chayanov did not deny that class differentiation could occur. However, Shanin's interpretation seems to preclude class differentiation (Cox 1986).

5. Harold Wolpe (1972) proposed the idea of articulation of modes of production to deal with the situation of different types of economy existing within one social system. His empirical analysis was of the Reserve system in South Africa as a subordinate, pre-capitalist mode based on kinship relations. Banaji (1977) offered a critique of Wolpe, who argued that the overall laws of motion of the economy define a social formation. Thus, it was not always necessary that all elements of the capitalist mode such as wage labour be present. For instance, petty commodity production or peasant production are forms of production *subordinate* to the capitalist mode (Banaji 1977: 19). Wolpe (1980: 36) accepted this and proposed the idea of 'extended' and 'restricted' modes.

6. Harriet Friedmann emphasised that simple commodity production is a form, not a mode, of production (1978, 1980). For her, simple commodity production is the unity of capital and labour within a household, contrary to the tendency within capitalism for the two to become separated. Petty commodity producing units are both enterprises and households. Friedmann put forward the 'double' specification thesis (cited in Llambi 1988: 361). Summarised, this is that simple commodity producing households are governed by external principles, mainly market conditions, as well as internal principles such as generational cycles, and the balance between consumption and investment. In practice, Chayanovian themes sometimes combine with those of Marxist derivation. Chevalier (1983) held that the category of simple commodity production embraces a large variety of forms within capitalism. Kahn argued that there is no absolute distinction between peasants and simple commodity producers (1982).

7. Low's (1986) analysis of farming systems in southern Africa does incorporate some gender dimensions. This is a neo-classical analysis of aspects of the region's agriculture such as the (assumed) lack of productivity of subsistence agriculture and the prevalence of labour migration. Moock's (1986) work on rational household strategies in African farming incorporates gender as a central dimension and points out how African systems differ from ordinary economic models. However, she departs far from Becker's model in order to make this sort of analysis, acknowledging that household members may have different and sometimes opposing goals.

8. Examining the term 'reproduction of labour power', Harris and Young (1981: 124) outline three different ways in which the reproduction of labour takes place under capitalism. These are:

   i)   the reproduction of individuals in specific class positions;
   ii) the reproduction of adequately socialised labour; and
   iii) the material reproduction of labour on an everyday basis.

9. More accurately, *some* groups of women may be more responsible for childcare. Low-status women may assume childcare responsibilities, often for higher-status ones.

10. In these struggles, ethnicity is stressed far more than gender divisions within indigenous communities (Deere and León 2000). To some extent, broaching of gender issues is seen as disruptive.
11. The number of structures emerges as an issue in several analyses. Pollert (1996) asks why Walby names six structures, rather than more or fewer. The same could be asked of Connell. However, Connell sees the theory as middle range, rather than definitive, so the choice seems as reasonable as any.

NOTES TO CHAPTER 3

1. These authors are concerned with a number of questions, including the effect of capitalist 'development' on gender relations; state policies; the effects of kinship relations on women's status and on policy outcomes; and with land reform issues. Deere and León have written about a range of Latin American nations, in particular Peru and Colombia (see Deere 1976, 1977, 1983; Deere and León 1982, 1987, 2001). They concentrate particularly on state policies, questions of capitalist development, and on class and gender relations among peasant smallholders. Since 1983, I have written about land reform and resettlement in Zimbabwe and southern Africa (Jacobs 1983). Agarwal's work on South Asia emphasises state policies, the impacts of technology, local land struggles, and the interaction of different (gendered) kinship systems with capitalism with regard to women's land rights (Agarwal 1985, 1994a, 1994b, 2003). Jean Davison has also discussed gender implications of land tenure and land reform in Mozambique and Malawi (1988a, 1988b, 1993). Manji (2003a, 2003b) examines gender and land reform in East Africa, especially with reference to World Bank policies. Many other writers too numerous to cite here, deal with gender and land issues, although not centrally with agrarian reform.
2. The countries included in case studies of household models of gender and land reform include: Burkina Faso (Upper Volta), Chile, Ethiopia, Honduras, India, Iran, Kenya, Libya, Nigeria, Peru, the Philippines, Poland, Sri Lanka, Tanzania, Viet Nam, and Zimbabwe. Chile and Peru are included here due to constraints of space, although detailed studies exist.
   Although this chapter concentrates upon reforms in which land is redistributed to households, it should be noted that the distinction between 'individual household' and collective reforms is not absolutely fixed. Many reforms involve both types of tenure, as in Chile.
3. Some, but not all, land was collectivised in Ethiopia's agrarian reform.
4. Women as mothers-in-law in patrilineal and patrilocal households often exert direct power over daughters-in-law. However, such power is *derived* from the spousal relation to the father-in-law (head of household) and so is not usually an example of autonomous exercise of power. Nevertheless, women in positions of partial power often actively oppress other women (see Jacobs 2000a, 2008).
5. As noted previously, much will depend upon contexts of culture, household structure, social class as well as individual families and personalities.

NOTES TO CHAPTER 4

1. A rare example is provided by the analysis of Spanish feminists in anarcho-syndicalist collectives during the Civil War. The *Mujeres Libres* put forward an analytical frame which differed from that of the mainstream

anarcho-syndicalist CGT (*Confederación General del Trabajo*). *Mujeres Libres* placed the sources of women's subordination not only in the economy but as having wider cultural roots as well (Ackelsburg 1993). Thus, their analysis prefigured much contemporary feminist thought, although it did not influence other collectivisation experiences.

2. Marketing cooperatives are very common across the world. However, as the name indicates, these are not production cooperatives and are not discussed here. Many production cooperatives also exist outside socialist economic trajectories.

3. In the 1920s, debates took place in the Soviet Union over the direction of economic development; these were to prove foundational in many respects for development theory. Bukharin advocated a strategy of gradual industrialisation and development of agriculture. Peasants should be persuaded not forced into collectives. Leeway should be given to middle peasants to cultivate and to market their goods. Preobrazhenskii, in contrast, advocated a much more rapid industrialisation, giving preference to industrial workers. Agriculture should be 'squeezed' and a collectivisation strategy pursued. This was seen as benefiting poorer peasants and agricultural workers. Eventually Stalin adopted the latter strategy. However, Preobrazhenskii had never advocated use of force against peasants and never envisaged the manner in which collectivisation would be carried out (Davies 1980).

4. The Stakhanovite movement began during the second Five Year Plan in 1935. This was named after Aleksei Stakhanov, who had mined fourteen times his quota of coal. The movement aimed to increase production, employing Taylorist efficiences, and was supported by the Communist Party (CPSU). It spread over other industries and into agriculture (Buckley 2006).

5. This proportion increased greatly during World War II and afterwards with the huge population imbalance that resulted from the death of millions of Soviet men.

6. From the early 1970s, administrative controls over internal migration, never completely effective, were relaxed (Zaslavskaya and Korel 1984).

NOTES TO CHAPTER 5

1. 'Chinese' usually refers to the entire population. However, it should be noted that one ethnic group, the Han, are numerically and culturally dominant. A number of other ethnic and religious minority groups exist.

2. Bossen (2002) cites Mao's *Report from Xunwu* published in 1960. An English translation was published by Stanford University Press in 1990: Mao Zedong, R.R. Thompson (translator) *Report from Xunwu*. Stanford Univ. Press.

3. Stacey (1983) cites Ramon Myers' *Chinese Peasant Economy* (1971) on the agrarian crisis in this period:

   i) Population growth led to increased landlessness; ii) a corrupt bureaucracy neglected the rural infrastructure characteristic of China (including irrigation and flood defences); and iii) the landlord class rack-rented the peasantry in an attempt to accumulate more land for itself. Dispossessed peasants supplied an almost inexhaustible supply of recruits for bands and warlord armies, contributing further to violence and chaos (Stacey 1983: 80). By the end of the nineteenth century, other developments also had profound effects. In the treaty ports, the Western enclave sector grew at the expense of Chinese enterprises so that cities were also exploiting the village economy. Traders, both Chinese and foreign, were able to control the sale of agricultural staples, reducing prices for peasant producers. The advent of manufactured goods also undermined the peasant handicraft economy. Floods and droughts, too, contributed greatly to the agricultural crisis. All of this resulted in forced land sales.

4. After 1962, many attempts to improve the grading of work points were made (Lin 1999).
5. Zhou (1996: 236) shows a bar chart but does not give precise figures.
6. One of the villages in Judd's study, Qianrulin, remained collective. People in the study noted that this was because they belonged to a single lineage (Judd 1994: 169).

   Evidence of explicit gender inequality in the Yunnan village included gender differences in basic allotment of land, contrary to national practice; provision that only men receive land allocations for public works and lack of provision for adjustment for women as they marry in.

NOTES TO CHAPTER 6

1. A *mu* is less than one-fifth of an acre
2. Some workers, especially single women, had broken away from their families in the 1980s to become integrated into the production collective
3. Kabeer and Tran's study indicates, however, that what is crucial for routes out of poverty appears to be diversification into *off-farm* activities, not diversification per se (2002: 183). Nevertheless, ordinary diversification is crucial for survival.
4. Korinek gives an example of Red River Delta women employed doing homework, weaving conical hats at exceedingly low rates of pay (2003: 188), much less than women earn in factory work.
5. Another FAO report (FAO 2005: 67) gives the figure for female-headed households as 32 per cent.

NOTES TO CHAPTER 7

1. *Machismo* (adj: *machista*) is usually seen as the cult or ideology of virile masculinity predominant in Latin America and the Caribbean. Machismo is a commonly used term and so can have different connotations, but it implies male domination within the family as well as control of female sexuality. Machismo is usually paired with marianist images sanctifying women in their roles as mothers. An ideal marianista woman is a pious and caring mother and housewife who is compliant and unassertive (Vargas-Lundius with Ypeij 2008). The gender regime denoted by these terms is usually seen as characteristic of Latin American societies.
2. However, if the ejidatario is absent for more than five years, whoever is in charge can now claim the land (Deere and León 2001: 152).
3. Some peasants were resettled without their consent, especially away from the Honduran border where contras had a base (Powelson and Stock 1987; Brunt 1995). The amount of expropriated land redistributed to peasant beneficiaries increased from 1985 to 1988 to 24 per cent of the total (Kay 2001: 759). Peasants also gained better access to scarce inputs

4. Montoya conducted interviews during the 1990s (Montoya 2003), but the study is included in this section as the research also refers to the Sandinista period.
5. Eighty per cent of onions, 75 per cent of beans, 62 per cent of pork, 54 per cent of corn, 50 per cent of grapes, and over 42 per cent of eggs and poultry are produced by the small farm sector in Brazil (Onsrud et al. 2005: 82).
6. The anti-dam movements began as movements for land, although later, in a different political climate, became reframed as ecological movements where they have gained greater prominence (Rothman and Oliver 1999).

NOTES TO CHAPTER 8

1. However, collectives were established in Mozambique under the previous FRELIMO (*Frente de Libertação de Moçambique*) government. In Tanzania, attempts were made to gather people together in collective ujamaa villages under Nyerere's African socialist policies. These were not successful in terms of production. (See Chapter 3 for reference to Tanzanian ujamaa villages.)
2. A number of matrilineal systems also exist in west and east/central Africa, although these are far fewer than patrilineal systems. They calculate lineage membership through a common ancestress and tend to give women more rights in land and property. Matriliny does not, however, equate to 'matriarchy' as a woman's maternal uncle or brother often exercises authority.
3. The honourific title 'widow' is also used to refer to divorcées.
4. This outlines the measurement of social class in my Zimbabwean research (Jacobs 1989). Two measures of social class (or strata within the peasantry formed through resettlement) were used. Both employed the measure termed 'self-sufficient': self-sufficiency in means of cultivation was defined as possession of two adult cattle and one plough.

   'Class 1': the variable Class 1 was calculated by addition of scores awarded to the following component variables:

   i)   does wage labour = 0; no wage labour = 1
   ii) no hiring in of labour = 0; hiring in of labour = 1
   iii) less than self-sufficient = 0; self-sufficient in means of cultivation = 1
   iv) cultivates 0 to 12 acres (ie the amount allocated) = 0; cultivates 13+ acres = 1

   'Class2': the variable Class 2 was calculated by the addition of a score for a fifth variable to 'Class 1', this concerning receipt of wage remittances:

   v)   Z\$0 to \$99 p.a. recieved in wage remittances = 0; receipt of Z\$100 + p.a. = 1.

   The stratum of 'poor peasants' consisted of peasants who were less than self-sufficient in their ability to cultivate. Despite having access to land since resettlement, they might lack sufficient draught animals or implements; many did wage labour for others, regularly or seasonally and were ex-commercial farm labourers.

   The 'wealthiest' stratum, on the other hand, had sufficient cattle and implements to plough; most hired in labour and one-fifth also managed to gain access to more than thirteen acres of land, due to (illegal) sharecropping arrangements. Many received wage remittances, usually from children, and these were important to household and individual income levels.

   The middle stratum was the largest, constituting nearly 50 per cent of the sample population in these northeastern RAs. A small majority were stereotypical 'middle peasants' in that they were independent of commodity production (they did not perform or hired in wage labour or receive remittances). The other half of the middle stratum was involved in complex sets of commodity relationships—for example both hiring in wage labour and performing wage labour themselves.

I used an individual measure of class so that I could disaggregate by sex. I constructed an indicator, or index, of 'power' or influence using answers to twenty-seven questions from the survey. The questions concerned matters such as wives' ability to make decisions or to influence them; the gender division of labour; behaviour and attitudes; and wives' perceptions of their own leverage within the home. Although less valid, in my view, than long-term observations, this index provided a useful if simplified measure. Using this calculation, it was found that a statistically significant relationship existed between 'Class 1' and the index 'wifepower'; the latter was divided into 'low' and 'high' for purposes of cross-tabulation. At the 5 per cent level of significance, with two degrees of freedom, the critical value of X^2 is 5.99. The observed value of X^2 was 6.11, indicating that a statistically significant relationship exists between the two variables.

Class 2 could be computed for only 77, rather then the full sample of 99 married women. At the 5 per cent level of significance, with two degrees of freedom, the critical value of X^2 is 5.99; at the 2.5 per cent level of significance, with 2 degrees of freedom, the critical value of X^2 is 7.38. The observed value was 6.27, indicating that a significant relationship exists between 'Class 2' and 'wifepower' for married women.

A statistically significant relationship also existed between 'husband's Class 2' and 'wifepower' for married women. At the 2.5 level of significance, with two degrees of freedom, the critical value of X^2 is 7.38. The observed value was 7.93, indicating that a significant relationship exists between the two variables.

5. Of the Zimbabwe Union of Democrats.
6. Matabeleland Operation *Gukurahundi* 1983 ('the early rain that washes away the chaff') consisted of a move by the North Korean-trained Fifth Brigade against the opposition party, ZAPU, led by Joshua Nkomo; this became translated into actions against Ndebele people in the west of the country. The estimates by the Catholic Commission for Justice and Peace report 'Breaking the Silence' were that approximately 20,000 Ndebele people were killed in this operation in 1983 and early 1984. A peace accord between ZANU and ZAPU was signed in 1987, which effectively absorbed the ZAPU party into ZANU-PF.
7. Ministry of Health and Child Welfare data from 2005 indicated a fall in adult HIV prevalence in Zimbabwe. AVERT (2008a) counsels that these data should be interpreted with caution, as it is not known whether the change indicates a temporary dip or is long term. An increase in deaths of HIV positive people due to conditions in the country, including malnutrition and lack of health care is thought to have contributed to the lowered rate of HIV infections. Additionally, like others, HIV positive people have migrated (AVERT 2008a).
8. Ian Scoones (Institute of Development Studies, Univ. of Sussex) and the Institute for Poverty, Land and Agrarian Studies (Univ. Western Cape) have carried out a study of Masvingo Province; this was published after this manuscript was drafted and so it has not been possible fully to incorporate the study's findings. The study is generally positive about the fast-track land reform programme in Zimbabwe (Scoones 2008).

9. Use of the SLAG grant and CPAs and trusts to access land appears to have been discouraged with the LRAD policy, although the grants are not officially discontinued (Hall 2007).

NOTES TO CHAPTER 9

1. Forward planning, sufficient inputs, extension advice, access to credit, and other assistance are all factors that help facilitate successful reforms.
2. The passing of gender-friendly land legislation is no simple matter, as evident in the Ugandan case. Here, provisions to give women equal rights to land were rescinded by the national government at the last minute despite a vigorous and initially successful national campaign (Kawamara-Mishambi and Ovonji-Odida 2003; Manji 2006).

Bibliography

Abdullah, Hussaina and Hamza, Ibrahim. (2003). 'Women and land in northern Nigeria: the need for independent ownership rights', in M. L. Wanyeki, (ed) *Women and Land in Africa*, London/Cape Town: Zed/D. Philip.

Ackelsburg, Martha. (1993). 'Models of revolution: rural women and anarchist collectivisation in civil war Spain', *J. Peasant Studies*, 203: 367–388.

Acker, Sandra. (1989). 'The problem with patriarchy', *Sociology*, 23(2): 235–240.

ActionAid. (2005). *Cultivating Women's Rights for Access to Land*, London: ActionAid. Online. Available HTTP: <www.fao.org/righttofood/KC/downloads/vl/docs/AH432.pdf> (accessed 28 November 2006).

Adam, David. (2006). 'Urban Population to Overtake Country Dwellers for First Time', *Guardian*, 16 June.

Adriance, Madeline. (1995). *Promised Land: base Christian communities and the struggle for the Amazon*, Albany: SUNY Press.

Agarwal, Bina. (1985). 'Work participation of rural women in the third world: some data and conceptual biases', *Economic and Political Weekly*, xx(51–52): 155–164.

———. (1989). 'Women, land and ideology in India', in H. Afshar and B. Agarwal (eds) *Women, Poverty and Ideology in Asia*, Basingstoke: Macmillan.

———. (1991). 'Agricultural mechanisation and labour use: a disaggregated approach', in H. Afshar (ed) *Women, Development and Survival in the Third World*, Harlow: Longmans.

———. (1994a). 'Gender and command over property: A critical gap in economic analysis and policy in South Asia', *World Development* 22(10): 1455–78.

———. (1994b). *A Field of One's Own: women and land rights in South Asia*, Cambridge, Cambridge University Press.

———. (2002). 'Disinherited peasants, disadvantaged workers: a gender perspective on land and livelihoods', in D. Thorner *Land, Labour and Rights: The Daniel Thorner Lectures*, Delhi: Tulika.

———. (2003). 'Gender and land rights revisited: exploring new prospects via the state, family and market', *J. Agrarian Change* 3(1–2): 184–224.

AGRITEX (Agricultural Extension Service of the Gov. of Zimbabwe). (1985). *Aspects of Resettlement*, Harare: Ministry of Agriculture.

Alavi, Hamza. (1973). 'Peasant classes and primordial loyalties', *J. Peasant Studies*: 1(1): 23–62.

———. (1982). 'Capitalism and the peasantry', in H. Alavi and T. Shanin (eds) *Introduction to the Sociology of Developing Societies*, London: Macmillan.

Alexander, Jocelyn. (1994). 'State, peasantry and resettlement in Zimbabwe', *Review of African Political Economy*, 21(61): 325–345.

Alexander, Sally and Taylor, B. (1980). 'In defence of patriarchy', *New Statesman*, 99, February 1.

All-China Women's Federation. (1991). 'The impact of economic development on rural women in China', in E. Masini and Statigos, S. (eds) *Women, Households and Change*, Tokyo: United Nations University: 181–204.

Allaghi, Farida. (1984). 'Rural women in a resettlement project: the case of the Libyan Arab Jamhiriya' in International Labour Organisation (ed) *Rural Development and Women in Africa*, Geneva: ILO: 137–45.

Althusser, Louis. (1971). 'Ideology and ideological state apparatuses' in Louis Althusser, *Lenin and Philosophy and Other Essays*, New York: Monthly Review.

Alvarez, Sonia. (1990). *Engendering Democracy in Brazil*, Princeton: Princeton University Press.

AMIHAN. (National Federation of Peasant Women–Philippines), Asian Peasant Women Network and Asia-Pacific Forum on Women, Law and Development. (2004). 'Asian peasant women conference on land rights & globalization': In AMIHAN Conference proceedings, Quezon City: University of the Philippines.

Amnesty International. (2002a). 'Zimbabwe: high risk of human rights violations', 14 March. Online. Available HTTP: <www.amnesty.org.uk/news_details. asp?NewsID=14207> (accessed 4 August 2008).

Amnesty International. (2002b). 'Zimbabwe: militias using food and sexual violence as tools of repression', 5 April. Online. Available HTTP: <www.amnesty. org.uk/news_details.asp?NewsID=14212> (accessed 4 August 2008).

———. (2007). 'Zimbabwe: women beaten and tortured for standing up to the government', 24 July. Online. Available HTTP: <www.amnesty.org.uk/news_ details.asp?NewsID=17414> (accessed 4 August 2008).

Amos, Valerie and Parmar, Prathiba. (1984). 'Challenging imperial feminism', *Feminist Review*, 17: 3–20.

Anderson, Perry. (1974). *Passages from Antiquity to Feudalism*, London: Verso.

Andors, Phyllis. (1976). 'Social revolution and woman's emancipation: China during the Great Leap Forward', *Signs*, 2(1): 33–42.

———. (1981). 'The 'Four Modernizations' and Chinese policy on women', *Bulletin of Concerned Asian Scholars* 13(2): 44–56.

Anthias, Floya. (2006). 'Concepts of intersectionality', paper presented at International Sociological Association World Congress: 'The quality of life in a globalising world' Durban, South Africa. 23–39 July

Anthias, Floya and Yuval-Davis, Nira. (1983). 'Contextualising feminism—gender, race and class divisions', *Feminist Review* 15: 62–74.

———. (1992). *Racialised Boundaries*, London: Routledge.

Appendini, Kirsten. (1996). 'Introduction to Panel II' in L. Randall (ed) *Reforming Mexico's Agrarian Reform*, Armonk, New York/London: M.E.Sharpe: 65–70.

Apthorpe, Raymond. (1981). 'A plea for a worm's eye view: policy oriented descriptive research on rural settlements and rural development', paper presented at University of East Anglia Regional Planning Conference: University of Cambridge, Cambridge. July

Arizpe, Lourdes and Botey, Carlota. (1987). 'Mexican agricultural development policy and its impact on rural women', in C. D. Deere and M. León de Leal (eds) *Rural Women and State Policy: feminist perspectives on Latin American agricultural development*, Boulder: Westview: 67–83.

Arrien, Juan. (2004). 'Literacy in Nicaragua', background paper prepared for The Education for All Global Monitoring Report. Online. Available HTTP: http:portal. unesco.org/education/en/file-download.php/ (accessed 12 December 2007).

Arrighi, Giovanni and Saul, John. (1973). *Essays on the Political Economy of Africa*, New York: Monthly Review Press.

Arruda, Roldão. (2004). 'Agrarian reform in Brazil: an interview with João Pedro Stédile'. Online. Available HTTP: <http://www.zmag.org/content/showarticle> (accessed 21 July 2006).

Arun, Shoba. (1999). 'Does land ownership make a difference? Women's roles in agriculture in Kerala, India', *Gender and Development* 7(3): 19–27.

Atemova, Olga. (2000). 'Changes in the everyday activities of rural women in Russia from the 1970s to the 1990s', in L. Alexander Norsworthy (ed) *Russian Views of the Transition in the Rural Sector,* Washington: World Bank: 132–44.

Aston, Basil; Hill, Kenneth, Piazza, Alan and Zeitz, Robin. (1984). 'Famine in China: 1958–62', *Population and Development Review,* 10(4): 613–645.

Asztalos Morrell, Ildikó. (1999). *Emancipation's Dead-End Roads? Studies in the formation and development of the Hungarian model for agriculture and gender (1956–89),* Studia Sociologica Upsaliensia 46; Uppsala: Acta Universitatis Upsaliensis.

AVERT (2008a) 'HIV and AIDS in Zimbabwe' 27/06/08. Online. Available HTTP: <http://www.avert.org/aids-zimbabwe.htm> (accessed 21 July 2008).

AVERT (2008b) 'HIV and AIDS statistics for South Africa' 09/6/08. Online. Available HTTP: <http://www.avert.org/safricastats.htm> (accessed 23 July 2008).

Banaji, Jarius. (1976). 'The peasantry in the feudal mode of production', *J. Peasant Studies,* 3(3): 299–320.

———. (1977). 'Modes of production in a materialist conception of history', *Capital and Class,* 3(1): 1–44.

———. (1990). 'Illusions about the peasantry: Karl Kautsky and the agrarian question', *J. Peasant Studies* 17(2): 288–307.

Banda, Fareda. (2005). *Women, Law and Human Rights: an African perspective,* Oxford, Hart.

Bandyopadhay, Rekha. (1996). 'Global review of land reform: a critical perspective', *Economic and Political Weekly,* 31 (11) 16 March: 679–91.

Banister, Judith. (2004). 'Shortage of girls in China today', *J. of Population Research,* 21(1): 19–45.

Barbic, Ana. (1993). 'Farm women in Slovenia: endeavours for equality', *Agriculture and Human Values,* 10(4): 3–25.

Barnard, Rosemary. (1983). 'Housewives and farmers: Malay women in the Muda irrigation scheme', in Lenore Manderson (ed) *Women's Work and Women's Roles,* Canberra: Australian National University: 129–145.

Barr, Abigail. (2004). 'Forging effective new communities: the evolution of civil society in Zimbabwean resettlement villages', *World Development* 32(10): 1753–1766.

Barraclough, Solon. (1991). *An End to Hunger?,* London: Zed Books.

Barrett, Michele and McIntosh, Mary. (1982). *The Anti-Social Family,* London: Verso.

Barrientos, Stephanie. (1997). 'The hidden ingredient: female labour in Chilean fruit exports', *Bulletin of Latin American Research:* 16(1): 71–81.

Barsted, Leila. (2005). 'Brazil: the legal status', in FAO *Gender and Land Compendium of Country Studies,* Geneva: Food and Agriculture Organisation of the United Nations. Online. Available FTP: <ftp://ftp.fao.org/docrep/fao/008/a0297e/a0297e00.pdf> (accessed 6 December 2006).

Baumeister, Eduardo. (2000). 'Institutional change and responses at the grassroots level: examples from Nicaragua, Honduras and El Salvador', in Annelies Zoomers and Gemma van der Haar (eds) *Current Land Policy in Latin America,* Amsterdam: KIT/Royal Tropical Institute: 249–258.

BBC (British Broadcasting Corporation). (2008). *Six o'clock News,* 8 July.

Beall, Jo. (2005). 'Decentralizing Government and Centralizing Gender in Southern Africa: lessons from the South African experience' UNRISD Occasional Paper 8, Geneva: UNRISD.

Becker, Gary. (1976). *The Economic Approach to Human Behaviour,* Chicago: Chicago University Press.

Bee, Anna. (2001). 'Agro-export production and agricultural communities: land tenure and social change in the Guatulame Valley, Chile', in A. Zoomers and G. van der Haar (eds) *Current Land Policy in Latin America*, Amsterdam: KIT/ Royal Tropical Institute: 17–28.

Bee, Anna and Vogel, Isabel. (1997). '*Temporeras* and household relations: seasonal employment in Chile's agro-export sector', *Bulletin of Latin American Research*, 16(1): 83–95.

Beechey, Veronica. (1978). 'Women and production: a critical analysis of some sociological theories of women's work', in Annette Kuhn and Anne-Marie Wolpe (eds) *Feminism and Materialism*, London: Routledge Kegan Paul.

———. (1979). 'On patriarchy', *Feminist Review*, 3: 66–82.

Bélanger, Danièle. (2002). 'Son preference in a rural village in North Vietnam', *Studies in Family Planning*, 33(4): 321–334.

Bélanger, Danièle. and Liu, Jianye. (2004). 'Social policy reforms and daughters' schooling in Vietnam', *International J. of Educational Development*, 24: 23–38.

Bell, Colin and Newby, Howard. (1976). 'Husbands and wives: dynamics of the deferential dialectic', in D. Barker and S. Allen (eds), *Dependence and Exploitation in Work and Marriage*, Harlow: Longmans Publishing.

Bennholdt-Thomsen, Veronika. (1981). 'Subsistence production and extended reproduction', in K. Young, C. Wolkowitz and R. McCullogh (eds) *Of Marriage and the Market*, London: Conference of Socialist Economists.

Bergmann, Barbara. (1995). 'Becker's theory of the family: preposterous conclusions', *Feminist Economics*, 1(1): 141–151.

Berkner, Lutz. (1972). 'Rural family organisation in Europe: a problem in comparative history', *Peasant Studies Newsletter*, 1: 145–56.

Bernstein, Henry. (1977). 'Notes on capital and peasantry', *Review of African Political Economy*, 10: 60–73.

———. (1979). 'Concepts for the analysis of contemporary peasantries', *J. Peasant Studies*, 6(4) : 421–43.

———. (1988a). 'Capitalism and petty bourgeois production: class relations and divisions of labour', *J. Peasant Studies*, 15(2): 258–71.

———. (1988b). 'Of virtuous peasants', in Shanin, Teodor (ed) *Peasants and Peasant Societies*, Harmondsworth: Penguin, 2nd edition.

———. (2001). "The peasantry' in global capitalism: who, where, and why?', in L. Panitch and C. Leys (eds) *The Socialist Register* 2001, New York: Merlin Press/ Fernwood Press/Monthly Review.

———. (2003). 'Land reform in southern Africa in world historical perspective', *Review of African Political Economy* 30(96): 203–226.

———. (2004). 'Changing before our very eyes: agrarian questions and the politics of land in capitalism today', *J. Agrarian Change* 4(3): 180–225.

Bernstein, Henry and Campbell, Bonnie. (1985). *Contradictions of Accumulation in Africa*, London: Sage.

Berry, Albert R. and Cline, William R. (1979). *Agrarian Structures and Productivity in Developing Countries*, Baltimore: Johns Hopkins University Press.

Besse, Susan. (1996). *Restructuring Patriarchy: the Modernization of Gender Inequality in Brazil: 1914–1940*, Chapel Hill: University of North Carolina Press.

Bich, Pham van. (1999). *The Vietnamese Family in Change: the case of the Red River Delta*, Richmond, Surrey: Nordic Institute of Asian Studies and Curzon.

Bikaako, Winnie and Ssemkumba, John. (2003). 'Gender, land and rights: contemporary contestations in law, policy and practice in Uganda', in L. Muthoni Wanyeki, (ed) *Women and Land in Africa*, London: Zed/Cape Town: D. Philip.

Binswanger, Hans and Elgin, Miranda. (1993). 'Reflections on land reform and farm size', in C. Eicher and J. Staatz (eds) *Agricultural Development in the Third World*, Baltimore: Johns Hopkins University Press.

Binswanger, Hans and Rosenzweig, Mark R. (1986). 'Behavioural and material determinants of production relations in agriculture', *J. Development Studies*: 22(3): 503–539.

Blumberg, Rae Lesser. (1988). 'Income under female vs. male control: hypothesies from theory and data from the third world', *J. Family Issues* 9(1): 51–84.

———. (1995). *Engendering Wealth and Wellbeing: empowerment for global change*, Boulder: Westview.

Bob, Urmilla. (1999). *African Rural Women and Land Reform in South Africa: case studies from the Midlands Region of KwaZulu Natal*, unpublished Ph.D thesis, W. Virginia University.

Bonate, Lizzat. (2003). 'Women's land rights in Mozambique' in L. Muthoni Wanyeki (ed) *Women and Land in Africa*, London: Zed/Cape Town: D. Philip.

Boni, Valdete. (2004). 'Poder e igualdade: as relações de gênero entre sindicalistas rurais de Chapecó, Santa Catarina' *Revista Estudos Feministas*, 12(1) 289–302 Online. Available HTTP: <http://www.scielo.br/scielo.php?script=sci_issuetoc&pid=0104–026X20040001> (accessed 21 December 2007).

Bonnell, Victoria. (1997). *Iconography of Power: soviet political posters under Lenin and Stalin*, Berkeley: University of California Press.

Borger, Julian. (2008). 'UN warns 5m Zimbabweans will face hunger by next year' *The Guardian* 19 June: 18.

Borras, Saturnino. (2003). 'Questioning market-led agrarian reform: experiences from Brazil, Colombia and South Africa', *J. Agrarian Change*, 3(3): 367–94.

———. (2005). 'Can redistributive land reform be achieved via market-based Land transfer schemes? Evidence and Lessons from the Philippines', *Journal of Development Studies*, 41(1): 90–134.

Borras, Saturnino. (2008). 'La Vía Campesina and its Global Campaign for Agrarian Reform' in S. Borras, Saturnino, M. Edelman, Marc and C. Kay (eds) *Transnational Agrarian Movements Confronting Globalization*, Chichester, Sussex: Wiley-Blackwell.

Borras, Saturnino, Edelman, Marc and Kay, Cristóbal (eds) (2008) *Transnational Agrarian Movements Confronting Globalization*, Chichester, Sussex: Wiley-Blackwell.

Boserup, Esther. (1970). *Women's Role in Economic Development*, New York: St. Martin's Press.

Bossen, Laurel. (2002). *Chinese Women and Rural Development: sixty years of change in Lu village*, Yunnan, Lanham, Md.: Rowman and Littlefield.

Bowyer-Bower, Tanya and Stoneman, Colin (eds). (2000). *Land Reform in Zimbabwe: constraints and prospects*, Aldershot: Ashgate.

Bradley, Harriet. (1989). *Men's Work, Women's Work*, Cambridge: Polity.

Brah, Avtar and Phoenix, Ann. (2004). 'Ain't I woman? revisiting intersectionality', *Journal of International Women's Studies*, 5(3): 75–86. Online. Available HTTP:www.bridgew.edu/SoAS/JIWS/May04/Phoenix_Brah.pdf(accessed 5 February 2007).

Branford, Sue and Glock, Oriel. (1985). *The Last Frontier: fighting over land in the Amazon*, London: Zed Books.

Bridger, Sue. (1996). 'The return of the family farm: a future for women?' in R. Marsh (ed) *Women in Russia and Ukraine*, Cambridge: Cambridge University Press: 241–54.

Broegaard, Rikke. (2005). 'Land tenure insecurity and inequality in Nicaragua' *Development and Change*, 36(5): 845–864.

Brown, Jennifer. (2004). Ejidos and Comunidades in Oaxaca, Mexico: impact of the 1992 reforms, Seattle: Rural Development Institute/RDI: RDI Reports on Foreign Aid and Development No.120, February. Online. Available HTTP: <http://www.rdiland.org/PDF/PDF_Reports/RDI_120.pdf> (accessed 30 January 2009).

Brownmiller, Susan. (1975). *Against Our Will*, New York: Simon and Schuster.

Bruce, John and Migot-Adhola, Shem. (1994). *Searching for Land Security in Africa*, Washington: World Bank.

Brumer, Anita. (2004). 'As agriculturas do sul do Brasil', *Revista Estudos Feministas*, 12(1): 205–227. Online. Available HTTP: <http://www.scielo.br/scielo.ph?script=sci_issuetoc&pid=0104–026X20040001> (accessed 6 December 2007).

Brunt, Dorien. (1992). *Mastering the Struggle: gender, actors and agrarian change in a Mexican ejido*, Amsterdam: CEDLA, the Netherlands.

———. (1995). 'Losing ground: Nicaraguan women and access to land during and after the Sandinista period' in CEDLA, *Agrarian Questions: the politics of farming*, Vol. I, Conference Proceedings, Wageningen: University of Wageningen: 273–90.

Bryceson, Deborah Fahy (ed) (1995) *Women Wielding the Hoe: lessons from rural Africa for feminist theory and development practice*, Oxford: Berg.

Bryceson, Deborah Fahy; Kay, Cristóbal and Mooij, Jos. (2000). *Disappearing Peasantries? Rural Labour in Africa, Asia and Latin America*, London: ITDG Books [Practical Action].

Buck, John L. (1957). *Land Utilisation in China*, New York: Council on Economic and Cultural Affairs.

Buckley, Mary. (2006). *Mobilizing Soviet Peasants: heroines and heroes of Stalin's fields*, Lanham, Md: Rowman and Littlefield.

Burawoy, Michael and Verdery, Katherine. (1999). 'Introduction' to M. Burawoy and K. Verdery (eds) *Uncertain Transition: ethnographies of change in the post-socialist world*, Lanham, Md: Rowman and Littlefield: 1–17.

Burawoy, Michael, Krotov, Pavel and Lytkina, Tatyana. (2000). 'Involution and destitution in capitalist Russia', *Ethnography:* 1, 1: 43–65.

Burbach, Roger. (1994). 'Roots of the postmodern rebellion in Chiapas', *New Left Review*, 205: 113–124.

Burger, Anna. (2001). 'Agricultural development and land concentration in a central European country: a case study of Hungary', *Land Use Policy*, 18: 259–68.

———. (2006). 'Why is the issue of land ownership still of major concern in East Central European (CEC) transitional countries, and particularly Hungary?', *Land Use Policy*, 23: 571–79.

Burgerné Gimes Anna and Szép Katalin. (2006). *Az egyéni (családi) mezőgazdasági helyzete napjainkban*, Budapest: Agroinform Kiadó.

Bush, Ray and Cliffe, Lionel. (1982). 'Labour migration and agrarian strategy in the transformation to socialism in southern Africa: Zimbabwe as a case', *Review of African Political Economy* conference, University of Liverpool.

Busingye, Harriett. (2002). 'Customary land tenure reform in Uganda: lessons for South Africa', Paper to PLAAS/Programme for Land and Agrarian Studies, Johanesburg. July

Business News [SA]. (2007). 'Harare sets aside paltry sum to pay out farmers' in *ZWNews*, 4 June. Online. Available HTTP: <http://zimbabwe-news.blogspot.com/2007_06_01_archive.html> (accessed 19 June 2008).

Butler, Judith. (1990). *Gender Trouble: feminism and the subversion of identity*, London: Routledge.

Bydawell, Moya. (1997). 'AFRA confronts gender issues: the process of creating a gender strategy', *Gender and Development* 5(1): 43–48.

Byers, Terry. (2004). 'Neo-classical neo-populism 25 years on: déjà-vu and déjà-passé]. *Agrarian Change*, 4 (1–2): 7–44.

Caldeira, Rute Rodrigues. (2006). *Disputed Meanings and Divergent Views: a study of MST leaders and settlers in Rio State, Brazil*, PhD thesis, Manchester: Manchester Metropolitan University, unpublished.

————. (2008). 'Updating its strategies and amplifying its frames: the Landless Workers' Movement in Brazil and the displacement of the struggle for land' *Journal of Perspectives on Global Development and Technology*, 7(2): 133–149.

Cameron, John; Jacobs, Susie, Ndhlovu, Tidings and Spring, Anita. (2006). 'The ILO's 'decent work' proposals: a concept paper', paper presented at International Sociological Association World Congress, 'The quality of life in a globalising world', Durban: 23–29 July.

Campbell, Connie and the Women's Group of Xapurí. (1996). 'Out on the Front Lines but still struggling for Voice: Women in the rubber tappers defense of the forest in Xapurí, Acre, Brazil.' D. Rocheleau, B. Thomas-Slayter and E. Wangari (eds), *Feminist Political Ecology*, London: Routledge.

Carr, Marilyn. (1991). *Women and Food Security: the African experience*, London: Intermediate Technology.

Casinader, Rex, Seplika Fernando and Karuna Gamage. (1987). 'Women's issues and men's roles: Sri Lankan village experience' in J. H. Momsen and J. Townsend (eds) *Geography of Gender in the Third World*, London: Hutchinson.

CDE (Centre for Development and Enterprise) (2008) 'Land reform in South Africa: getting back on track', 16 May. Online. Available HTTP: <http://www.cde.org.za/article.php?a_id=284> (accessed 30 July 2008).

Ceci, Sara. (2005). 'Nicaragua: access to land', in Food and Agriculture Organisation of the United Nations (ed) *Gender and Land compendium of country studies*, Geneva: Food and Agriculture Organisation of the United Nations. Online. Available FTP: <ftp://ftp.fao.org/docrep/fao/008/a0297e/a0297e00.pdf> (accessed 6 December 2006).

Chakrabarti, Anil (2003): *Beneficiaries of Land Reforms: the West Bengal scenario*. State Institute of Panchayats and Rural Development, Kalyani.

Chaliand, Gérard. (1969). *The Peasants of North Vietnam*, Harmondsworth: Penguin.

Chambers, Robert. (1988). Personal communication, London.

Chanock, Martin. (1989). 'Neither customary nor legal: African customary law in an era of family law reform', *International Journal of Law, Policy and the Family*, 3(1): 72–88.

Chant, Sylvia. (2004). 'Dangerous equations? how female-headed households became the poorest of the poor: causes, consequences and cautions', *IDS Bulletin*, 35(4): 19–26.

Chant, Sylvia and Campling, Jo. (1997). *Woman-Headed Households*, Houndmills: Macmillan.

Chasin, Barbara. (1990). 'Land reform and women's work in a Kerala village', Women in Development Working Paper 207: Lansing: Michigan State University.

Chayanov, A.V. (1989). *Theory of Peasant Economy*, Manchester: Manchester University Press.

Cheater, Angela. (1981). 'Women and their participation in commercial agricultural production', *Development and Change*, 12(July): 349–77.

Chenaux-Répond, Maia. (1994). 'Gender-Based Land Use-Rights in Model A Resettlement Schemes of Mashonaland, Zimbabwe', ZWCN Monographs, Harare.

Chevalier, Jacques. (1983). 'There is nothing simple about simple commodity production', *J. Peasant Studies* 10(4): 153–86

Chikondo, Knowledge. (1996). 'Production and Management of Natural Resources in Resettlement Areas in Zimbabwe: the case of Msasa-Ringa', Agricultural University of Norway, MSc dissertation, unpublished. (cited in Kinsey 2004b).

Chimedza, Ruvimbo. (1988). Women's Access to and Control over Land: the case of Zimbabwe, Department of Agricultural Economics, Working Paper 10/88, Harare: University of Zimbabwe.

Chimhowu, Admos and Hulme, David. (2006). 'Livelihood dynamics in planned and spontaneous resettlement in Zimbabwe: converging and vulnerable', *World Development*, 34(4): 728–750.

Christodoulou, Dimitrious. (1990). *The Unpromised Land: agrarian reform and conflict worldwide*, London: Zed Books.

CIA (Central Intelligence Agency of the United States of America) (2006) 'Nicaragua: country facts'. Online. Available HTTP: <https://www.cia.gov/library/publications/the-world-factbook/index.html> (accessed 3 January 2007).

Claassens, Aninke. (2003). *Community Views of the Communal Land Rights Bill*, Research Report no. 15, Programme for Land and Agrarian Studies, University of the Western Cape.

Cliffe, Lionel. (2000). 'The politics of land reform in Zimbabwe' in T. Bowyer-Bower, and C. Stoneman (eds) *Land reform in Zimbabwe: constraints and prospects*, Aldershot: Ashgate: 35–57.

Colburn, Forrest. (1986). *Post-Revolutionary Nicaragua: state, class, and the dilemmas of agrarian policy*, Berkeley: University of California Press.

Collinson, Helen. (1990). *Women and Revolution in Nicaragua*, London: Zed Books.

Colson, Elizabeth. (1960). *The Social Organisation of the Gwembe Tonga*: Vol. I *The Human Consequences of Resettlement*, Lusaka: University of Zambia.

Commission for Gender Equality. (2003). Interview with anonymous informant by S. Jacobs, Johannesburg.

Connell, R. W. (1987). *Gender and Power*, London: Allen and Unwin.

Conti, Anna. (1979). 'Capitalist organisation of production through non-capitalist relations: women's role in a pilot resettlement, Upper Volta', *Review of African Political Economy*, 15–16: 75–92.

Cornwall, Andrea, Harrison, Elizabeth and Whitehead, Ann. (2006*). Gender Myths and Feminist Fables: the struggle for interpretive power in gender and development*, Oxford: Blackwell.

Cousins, Benjamin. (1997). 'How do rights become real? Formal and informal institutions in South Africa's land reform', *IDS Bulletin*, 28(4): 59–67.

——. (2006). 'Debating agrarian reform in South Africa: responses to an article' in *Mail & Guardian*, 8 August 2006 (e-mail correspondence).

——. (2007). 'Agrarian reform and the 'two economies': transforming South Africa's countryside', in Ntsebzah, L. and Hall, R. (eds) *The Land Question in South Africa*, Cape Town: HSRC.

Cox, Terence. (1986). *Peasants, Class and Capitalism*, Oxford: Clarendon/Oxford University Press.

Croll, Elisabeth. (1980). *Feminism and Socialism in China*, London: Routledge.

——. (1981). *The Politics of Marriage in Contemporary China*, Cambridge: Cambridge University Press.

——. (1985). 'Fertility norms and family size in China', in E. Croll, D. Davin and P. Kane (eds) *China's One Child Family Policy*, New York: St. Martin's Press.

——. (1987a). 'New peasant family forms in rural China', *J. Peasant Studies*, 14(4): 469–99.

——. (1987b). 'Some implications of rural economic reforms for the Chinese peasant household', in A. Saith (ed) *The Re-emergence of the Chinese Peasantry*, Beckenham, Kent: Croom Helm.

Cromwell, Elizabeth and Chintedza, Allan. (2005). 'Neo-patrimonialism and policy processes: lessons from the Southern African food crisis', *IDS Bulletin* 36(2): 103–108.

Crook, David and Crook, Isabel. (1979). *Revolution in a Chinese Village: Ten Mile Inn*, NY: Random House.

Cross, Catherine. (1992). 'An alternate legality: the property rights question in relation to South African land reform', *S. African Journal on Human Rights*, 305–31.

Cross, Catherine and Friedman, Michelle. (1997). 'Women and land: marginality and the left-hand power', in S. Meer (ed) *Women, Land and Authority*, Oxford, Oxfam: 17–34.

Cusworth, John. (2000). 'A review of the UK ODA evaluation of the Land Resettlement Programme in 1988 and the Land Appraisal Mission of 1996', in T. A. S. Bowyer-Bower and C. Stoneman (ed) *Land Reform in Zimbabwe: constraints and prospects*, Aldershot: Ashgate: 25–34.

Dao, The Ruan. (1995). 'The peasant household and social change', in B.J. Tria Kerklievt and D. Porter (eds) *Vietnam's Rural Transformation*, Boulder: Westview.

Davin, Delia. (1976). *Woman-Work*, Oxford: Oxford University Press.

——. (1987). 'China—the new inheritance law and the peasant household', *J. Communist Studies*, 3(4): 52–63.

——. (1988). 'The Implications of contract agriculture for the employment and status of Chinese peasant women', in S. Feuchtwang, A. Hussain and T. Pairault *Transforming China's Economy in the Eighties*, Vol. I, London: Zed.

Davis, Benjamin, Carletto, Calogero and Piccione, Norman. (2001). 'Income generation strategies among Nicaraguan agricultural producers', in Annelies Zoomers and Gemma van der Haar (eds) *Current Land Policy in Latin America*, Amsterdam: KIT/Royal Tropical Institute: 169–190.

Davies, Robert W. (1980). *The Socialist Offensive: the collectivisation of Soviet agriculture: 1929–30*, London: Macmillan.

Davison, Jean (1988a) 'Land and women's agricultural production: the context', in J. Davison (ed) *Women, Agriculture and Land: the African experience*, Boulder: Westview.

——. (1988b). 'Land redistribution in Mozambique and its effects on women's collective production: case studies from Sofala Province', in J. Davison (ed) *Women, Agriculture and Land: the African experience*, Boulder: Westview: 228–49.

——. (ed) (1988) *Women, Agriculture and Land: the African experience*, Boulder: Westview.

——. (1993). 'Clinging to banja production in face of changing gender relations in Malawi', *J. Southern African Studies* 19(3): 405–421.

Davydova, Irina and Franks, J.R. (2006). 'Responses to agrarian reforms in Russia: evidence from Novosibirsk oblast', *Journal of Rural Studies*, 22(1): 39–54.

Dawson, Alexander. (2006). *First World Dreams: Mexico since 1989*, London: Zed Books

Day, Shelagh and Brodsky, Gwen. (1998). '*Women and the Equality Deficit: the impact of restructuring Canada's social programs*', Ottawa: Status of Women, Canada.

De Walt, Billie, Rees, Martha with Murray, Arthur. (1994). *Past Lessons, Future Prospects: the end of the agrarian reform in Mexico*, San Diego: Centre for US-Mexican Studies.

Deere, Carmen Diana. (1976). 'Rural women's subsistence production in the capitalist periphery', *Review of Radical Political Economics*, 8(1): 9–17

——. (1977). 'Changing social relations of production and Peruvian peasant women's work', *Latin American Perspectives*, 4(1–2): 48–69.

——. (1983). 'Cooperative development and women's participation in the Nicaraguan agrarian reform', *American J. of Agricultural Economics* (December): 1043–48.

——. (1986a). 'Agrarian reforms, peasant participation and the organisation of production in the transition to socialism', in R. Fagen, C. Deere and J. Corragio (eds) *Transition and Development: problems of third world socialism*, NY: Monthly Review Press.

——. (1986b). 'Rural women and agrarian reform in Peru, Chile and Cuba', in J. Nash and H. Safa (eds) *Women and Change in Latin America*, S. Hadley, Mass: Bergin and Garvey.

——. (1987). 'The Latin American agrarian reform experience', in C.D. Deere and M. León de Leal (eds) *Rural Women and State Policy: feminist perspectives on Latin America agricultural development*, Boulder: Westview.

——. (1997). 'Reforming Cuban agriculture', *Development and Change*, 28: 649–69.

——. (with the Rural Studies Team, Univ. of Havana) (1998). 'Cuba: the reluctant reformer', in I. Szélenyi (ed) *Privatizing the Land*, London: Routledge: 62–91.

——. (2003) 'Women's land rights and social movements in the Brazilian agrarian reform', *J. Agrarian Change*, 3, 1–2.

Deere, Carmen Diana and León de Leal, Magdalena (1982). *Women in Andean Agriculture*, Geneva: ILO.

——. (eds) (1987). 'Introduction', in C. D. Deere and M. León (eds) *Rural Women and State Policy: feminist perspectives on Latin America agricultural development*, Boulder: Westview: 1–17.

——. (2000). 'Neo-liberal agrarian legislation, gender equality and indigenous rights: the impact of new social movements' in A. Zoomers and G. van der Haar (eds) *Current Land Policy in Latin America*, Amsterdam: KIT/the Royal Tropical Institute: 75–92.

——. (2001). *Empowering Women: land and property rights in Latin America*, Pittsburgh: Pittsburgh University Press.

——. (2003). 'The gender asset gap: land in Latin America', *World Development*: 31(6): 925–47.

Deere, Carmen Diana and Pérez, Niurka. (1999). 'Cuba: successful voluntary collectivization', in Mieke Meurs (ed) *Many Shades of Red*, Lanham, Md: Rowman and Littlefield.

Dekker, Henri. (2003). *The Invisible Line: land reform, land tenure security and land registration*, Aldershot: Ashgate.

Delphy, Christine. (1977). *The Main Enemy*, London: Women's Research and Resource Centre.

Diamond, Norma. (1975). 'Collectivization, kinship and the status of women in rural China', *Bulletin of Concerned Asian Scholars*, 7 (1): 25–32.

Disney, Jennifer. (2004). 'Incomplete revolutions: gendered participation in productive and reproductive labor in Mozambique and Nicaragua', *Socialism and Democracy*, 18, 1(35): 7–42.

Dixon-Mueller, Ruth. (1985). *Women's Work in Third World Agriculture*, Geneva: International Labour Office.

Dolny, Helena. (1985). 'The challenge of agriculture', in J. Saul (ed) *The Difficult Road to Socialism in Mozambique*, New York: Monthly Review Press.

Dorner, Peter and Thiesenhausen, Wm. (1990). 'Selected land reforms in East and Southeast Asia: their origins and impacts', *Asian-Pacific Economic Literature*, 4(1): 65–95.

Duncan, Jennifer and Li Peng. (2001). *Women and Land Tenure in China: a study of women's land rights in Dongfang County, Hainan*, Seattle: Rural Development Institute. Online. Available HTTP: <http://www.rdiland.org/PDF/PDF_Reports/RDI_110.pdf> (accessed 10 October 2006).

Dwyer, Daisy and Bruce, Judith (eds) (1988) *A Home Divided: women's income in the third world*, Stanford: Stanford University Press.

Dyer, Graham. (2004). 'Redistributive land reform: no April rose. the poverty of Berry and Kline and GKI on the inverse relationship', *J. Agrarian Change*, 4(1–2): 45–72.

Edholm, Felicity, Harris, Olivia and Young, Kate. (1977). 'Conceptualising women', *Critique of Anthropology*, 3(9–10): 101–30.

Einhorn, Barbara. (1993). *Cinderella Goes to Market*, London: Verso.

Eisen, Arlene. (1984). *Women and Revolution in Vietnam*, London: Zed Books.

Eisenstein, Zillah. (1979). *Capitalist Patriarchy and the Case for Socialist Feminism*, New York: Monthly Review Press.

El-Ghonemy, M. Riad. (1990). *The Political Economy of Rural Poverty: the case for land reform*, London: Macmillan.

———. (2001). 'The political economy of market-based land reform', in Ghimire, K. (ed) *Land Reform and Peasant Livelihoods*, London: ITDG.

Ellis, Frank. (1988). *Peasant Economics*, Cambridge: Cambridge University Press.

———. (2000). *Rural Livelihoods and Diversity in Developing Countries*, Oxford: Oxford University Press.

Elson, Diane (ed) (1995) *Male Bias in the Development Process*, Manchester: Manchester University Press, 2nd edition.

———. (2002). 'Gender justice, human rights and neo-liberal economic policies', in M. Molyneux and S. Ravazi (eds) *Gender Justice, Development and Rights*, Oxford: Oxford University Press: 78–114.

Elson, Diane and Gideon, Jasmine. (2004). 'Organising for women's economic and social rights: how useful is the International Covenant on Economic, Social and Cultural Rights?', *Journal of Interdisciplinary Gender Studies*: 8(1–2): 133–52.

Engels, Frederick. (1972). *Origins of the Family, Private Property and the State*, London: Lawrence and Wishart.

Ennew, Judith; Hirst, Paul and Tribe, Keith. (1977). '"Peasantry' as an economic category', *J. Peasant Studies*, 4(4): 295–322.

Entwisle, Barbara and Henderson, Gail. (2000). 'Conclusion: re-drawing boundaries', in B. Entwisle and G. Henderson (eds) *Re-Drawing Boundaries: work, household and gender in China*, Berkeley: University of California Press: 284–304.

Entwisle, Barbara; Short, Susan; Fengying, Zhai and Linmao, Ma. (2000). 'Household economies in transitional times', in Barbara Entwisle and Gail Henderson (eds) *Re-Drawing Boundaries: work, household and gender in China*, Berkeley: University of California Press: 261–83.

Eschle, Catherine. (2001). *Global Democracy, Social Movements and Feminism*, Boulder: Westview.

Evans, Allison. (1991). 'Gender issues in rural household economics', *IDS Bulletin*, 22 (1): 51–9.

———. (1992). 'Statistics', in Lise Østergaard (ed) *Gender and Development: a practical guide*, London: Routledge.

Evans, Mary. (1982). *The Woman Question; readings on the subordination of women*, London: Fontana.

Fallers, Lloyd. (1961). 'Are African cultivators to be called peasants?', *Current Anthropology*, 2(2): 108110.

Faludi, Susan. (1992). *Backlash: the undeclared war on American women*, London: Chatto and Windus.

FAO (Food and Agriculture Organisation of the United Nations) (2000) 'Contemporary thinking on land reform'. Online. Available HTTP: <http://www.caledonia.org.uk/land/fao.htm> (accessed 20 February, 2006).

———. (2002). '*Gender Differences in the Transitional Economy of Viet Nam: household structure and living standards*', Ha Noi, Viet Nam: FAO and UNDP. Online. Available HTTP: <http://www.fao. /docrep/005/ac685e/ac685e05.htm> (retrieved 31 January 2009).

———. (2005). *Gender and Land Compendium of Country Studies*, Geneva: Food and Agriculture Organisation of the United Nations. Online. Available

FTP: <ftp://ftp.fao.org/docrep/fao/008/a0297e/a0297e00.pdf> (accessed 15 August 2006).

Fapohunda, Eleanor. (1987). 'The nuclear household model in Nigeria', *Development and Change*, 18(2): 281–91.

Farnsworth, Beatrice and Viola, Lynne. (1992). *Russian Peasant Women*, Oxford: Oxford University Press.

Fei, Hsiao-Tung (1939). *Peasant Life in China: a Field Study of Country Life in the Yamgtze Valley*, London: Routledge.

Feng, Wang. (2000). 'Gendered migration and the migration of genders', in Barbara Entwisle and Gail Henderson (eds) *Re-Drawing Boundaries: work, household and gender in China*, Berkeley: University of California Press: 231–42.

Ferree, Myra Marx and Tripp, Aili Marie (eds). (2006). *Transnational Women's Activism, Organizing and Human Rights*, New York: New York University Press.

Ferret, Grant. (2004). 'Zimbabwe farms 'left unsettled'', British Broadcasting Coorporation. Online. Available HTTP: <http://news.bbc.co.uk/1/hi/world/africa/3364267.stm> (accessed 4 August 2008).

FIAN (Face It Act Now: Fighting Hunger with Human Rights) (2004) leaflet, unpublished, distributed at FMRA World Congress, Valencia, Spain.

Figueroa Albelo, Victor. (2003). 'Cuba: an experience of rural development', in V.K. Ramachandran and Madhura Swaminathan (eds) *Agrarian Studies: essays on agrarian relations in less-developed countries*, London: Zed Books: 506–31.

Figueroa Albelo, Victor and A. Averhoff Casamayor. (2001). 'La agricultura cubana y la reforma agraria de 1993', *Land reform, land settlement and cooperatives*. FAO Corporate Document Depository. Online. Available HTTP: http://www.fao.org/DOCREP/005/Y2519T/y2519t06.htm (accessed 25 June 2009). Available at ftp://ftp.fao.org/docrep/ fao/004/y2519T/y2519Tpdf, accessed 15 Nov. 2006.

Firth, Raymond. (1966). *Malay Fishermen: their peasant economy*, London: Routledge.

Fitzpatrick, Sheila. (1994). *Stalin's Peasants*, Oxford: Oxford University Press.

Floro, Maria. (1992). 'Women's work and agricultural commercialization in the Philippines', in N. Folbre, B. Bergmann, B. Agarwal and M. Floro (eds) *Women's Work in the World Economy*, Oxford: Oxford University Press.

FMRA (Foro Mundial sobre la Reforma Agraria). (2004). 'Statement: for a world without hunger'. Online. Available HTTP: <http://www.fmra.org/declaracion_final.doc> (accessed 31 January 2005).

Folbre, Nancy. (1984). 'Household production in the Philippines: a non-neo-classical approach', *Economic Development and Cultural Change*, 32(2): 303–330.

———. (1986a). 'Hearts and spades: paradigms of household economics', *World Development*, 14(2): 245–55.

———. (1986b). 'Cleaning house: new perspectives on households and economic development' *J. Development Economics*, 22: 5–40.

———. (1988). 'Patriarchal social formations in Zimbabwe', in S. Stichter and Jane Parpart (eds) *African Women in the Household and the Workplace*, Beverley Hills: Sage.

Fortin, Elizabeth. (2005). 'Reforming land rights: the World Bank and the globalisation of agriculture', *Social and Legal Studies*, 14(2): 147–177.

Foweraker, Joe. (1981). *The Struggle for Land: a political economy of the pioneer frontier in Brazil from 1930 to the present day*, Cambridge: Cambridge University Press.

Francis, Elizabeth. (2000). *Making a Living: rural livelihoods in Africa*, London: Routledge.

Freedman, Maurice. (1970). *Lineage Organization in Southeastern China*, London: Athlone Press, 3rd edition.

——. (1971). *Chinese Lineage and Society: Fukien and Kwangtung*, London: Athlone Press.

Frenier, Mariam Darce. (1983). 'The effects of the Chinese Communist land reform on women and their families', *Women's Studies International Forum*, 6(1): 41–55.

Friedmann, Harriet. (1978). 'World market, state and family farm: social relations of household production in an era of wage labor', *Comparative Studies in Society and History*, 20(4): 545–86.

——. (1980). 'Household production and the national economy: concepts for the analysis of agrarian formations', *J. Peasant Studies*, 7(2): 158–84.

——. (1986a). 'Patriarchal commodity production', *Social Analysis*, 20: 47–55.

——. (1986b). 'Patriarchy and property—a reply to Goodman and Redclift', *Sociologia Ruralis*, XXVI (2): 186–93.

——. (1986c). 'Family enterprises in agriculture: structural limits and political possibilities', in G. Cox, P. Lowe and M. Winter (eds) *Agriculture: people and policies*, London: Allen and Unwin.

Gaidzanwa, Rudo. (1995). 'Land and the economic empowerment of women: a gendered analysis', *SAFERE*, 1(1). 1–12.

Gammeltoft, Tine. (1999). *Women's Bodies, Women's Worries: health and family planning in a Vietnamese rural community*; Richmond, Surrey: Nordic Institute of Asian Studies.

García-Frias, Zoraida. (2005). 'Introduction', in FAO *Gender and Land Compendium of Country Studies*, Geneva: Food and Agriculture Organisation of the United Nations. Online. Available FTP: <ftp://ftp.fao.org/docrep/fao/008/a0297e/a0297e00.pdf> (accessed 15 July 2006).

Gardiner, Jean. (1975). 'Women's domestic labour', *New Left Review*, 89: 47–58.

Garrett, Patricia. (1982). 'Women and agrarian reform: Chile, 1964–73', *Sociologia Ruralis*, 22(1): 17–29.

Gawaya, Rose. (2008). 'Investing in women farmers to eliminate food insecurity in southern Africa: policy-related research from Mozambique', *Gender and Development*, 6(1): 147–59

Ghimire, Krishna (ed). (2001). *Land Reform and Peasant Livelihoods*, London: ITDG.

Giani, Luiz; de Souza, Maria de Fátima; Mager, Miryam and da Silva, Renata (2003). 'Discriminação de gênero e participação política: a atuaço da mulher sem terra', *Acta Scientarum*, 25(1): 159–64.

Gibbon, Peter and Neocosmos, Michael. (1985). 'Some problems in the political economy of 'African Socialism'', in H. Bernstein and B. Campbell (eds) *Contradictions of Accumulation in Africa*, Bevely Hills: Sage.

Glavanis, K. (1989). 'The small peasant household in Egypt' in K and Pandeli Glavanis (eds) *The Rural Middle East*, London: Zed.

Gledhill, John. (1991). *Casi Nada: a study of agrarian reform in the homeland of Cardenismo*: Studies on Culture and Society, volume 4: Albany, NY: Institute for Mesoamerican Studies, University of New York at Albany/University of Texas Press.

Glickman, Rose. (1992). 'Peasant women and their work', in B. Farnsworth and L. Viola (eds) *Russian Peasant Women*, Oxford: Oxford University Press.

Global Exchange. (2005). Factsheet. Online. Available HTTP: <http://www.globalexchange.org/countries/americas/mexico/factsheet.pdf> (accessed 2 February 2009).

Globalis. (2005). South Africa: urban population, *Globalis*. Online. Available HTTP: <http://globalis.gvu.unu.edu/indicator_detail.cfm?IndicatorID=30&Country=ZA> (accessed 29 July 2008).

Globalis. (2008). 'Brazil: urban population', *Globalis*. Online. Available HTTP: <http://globalis.gvu.unu.edu/indicator_detail.cfm?IndicatorID=30&Country=BR> (accessed 4 December 2008).

Glucksmann, Miriam. (2005). 'Shifting boundaries and interconnections: extending the 'total social organisation of labour", in L. Pettinger, J. Parry, R. Taylor, and M. Glucksmann (eds) *A New Sociology of Work?*, Oxford: Blackwell Publishers.

Goebel, Allison. (1998). 'Process, participation and power: notes from 'participatory' research in a Zimbabwean resettlement area', *Development and Change*, 29: 277–307.

———. (1999). "Here it is our land, the two of us': women, men and land in a Zimbabwean resettlement area', *J. Contemporary African Studies*, 17: 75–96.

———. (2005a). *Gender and Land Reform: the Zimbabwean experience*, Montreal: McGill-Queens University Press.

———. (2005b). 'Zimbabwe's 'fast-track' land reform: what about women?', *Gender, Place and Culture* 12(2): 145–72.

Goetz, Anne-Marie and Hassim, Shireen (eds) (2003) *No Shortcuts to Power: African women in politics and policy-making*, London: Zed: 160–187.

Goheen, Miriam. (1988). 'Land and the household economy: women of the grassfields today', in J. Davison (ed) *Women, Agriculture and Land*, Boulder: Westview.

Goldring, Luin. (1996). 'The changing configuraton of property rights under ejido reform', in Laura Randall (ed) *Reforming Mexico's Agrarian Reform*, Armonk, New York/London: M.E.Sharpe: 271–288.

Goldschmidt-Clermont, L. (1982). *Unpaid Work in the Household*, Geneva: International Labour Organisation.

Goldstein, Sidney, Zai Liang and Goldstein, Alice. (2000). 'Migration, gender and the labor force in Hubei Province, 1985–1990', in Barbara Entwisle and Gail Henderson (eds) *Re-Drawing Boundaries: work, household and gender in China*, Berkeley: University of California Press: 214–30.

Goode, William. (1963). *World Revolution and Family Patterns*, New York: Free Press.

Goodman, David and Redclift, Michael. (1981). *From Peasant to Proletarian*, Oxford: Blackwell.

Gouthami and Rajgor, Meena. (2008). 'Women's perceptions of land ownership: a case study from Kutch District, Gujarat, India', *Gender and Development*, 16(1): 41–54.

GOZ (Government of Zimbabwe). (1985). *Zimbabwe, Intensive Resettlement Programme: policies and procedures*, Harare: Department of Rural Development.

GOZ (Government of Zimbabwe). (2003). 'Report of the Presidential Land Review Committee on the Implementation of Fast-Track Land Reform: 2000–2002', under the chairmanship of Dr. Charles Utete: Harare ('Utete Report').

Gray, Jack. (1990). *Rebellions and Revolutions: China from the 1800s to the 1980s*, Oxford: Oxford University Press.

Green, Linda. (1996). 'What's at stake? The reform of agrarian reform in Mexico', in Laura Randall (ed) *Reforming Mexico's Agrarian Reform*, Armonk, New York/London: M.E. Sharpe: 267–270.

Gregson, N and Foord, J. (1987). 'Patriarchy—comments on critics', *Antipode*, 19(3): 371–75.

Griffin, Keith; Kahn, A.R. and Ickowitz, Amy. (2002). 'Poverty and the distribution of land', *J. Agrarian Change*, 2(3): 279–330.

———. (2004). 'In defence of neo-classical neo-populism', *J. Agrarian Change*, 4(3): 361–86.

Grindle, Merilee. (1996). 'Introduction to Panel III' in Laura Randall (ed) *Reforming Mexico's Agrarian Reform*, Armonk, New York/London: M.E. Sharpe: 145–50.

Groenewald, Yolandi. (2008). 'Land reform hobbled by capacity problems', *Mail and Guardian online*, 22/2. Online. Available HTTP: <http://www.mg.co.za/article/2008-02-22-land-reform-hobbled-by-capacity-problems> (accessed 27 July 2008).

Gu, Jiantang. (1988). 'Marriage and the family', in China Financial and Economic Publishing House, *New China's Population*. New York: Macmillan: 129–49.

Guivant, Julia. (2003). Agrarian Change, Gender and Land Rights: a Brazilian case study, Geneva: UNRISD Social Policy and Development Paper no. 14.

Guyer, Jane. (1981). 'Household and community in Africa', *African Studies Review*, XXIV (2–3): 87–137.

———. (1986). 'Intra-household processes and farming systems research: perspectives from anthropology', in J. Moock (eds) (1986) *Understanding Africa's Rural Households and Farming Systems*, Boulder: Westview.

Guyer, Jane and Peters, Pauline. (1987). 'Introduction', *Development and Change*: special issue on *conceptualising the household—issues of theory and policy in Africa*: 18(2): 197–214.

van der Haar, Gemma. (2000). 'The 'Indianization' of land reform: the Tojolabal Highlands of Chiapas, Mexico', in Annelies Zoomers and Gemma van der Haar (eds) *Current Land Policy in Latin America*, Amsterdam: KIT/Royal Tropical Institute: 147–160.

Hale, Angela and Wills, Jane. (2007). 'Women Working Worldwide: transnational networks, corporate social responsibility and action research', *Global Networks* 7(4): 453–76.

Hall, Anthony. (1990). 'Land tenure and land reform in Brazil', in Roy Prosterman, Mary Temple and Timothy Hanstad (eds) *Agrarian Reform and Grassroots Development*, Boulder: Lynne Rienner.

Hall, Ruth. (2004). 'Land and agrarian reform in South Africa: a status report 2004', Research Report no. 15, Programme for Land and Agrarian Studies, University of the Western Cape.

———. (2007). 'Transforming rural South Africa? Taking stock of land reform', in Lungisile Ntsebeza. and Ruth Hall (eds) *The Land Question in South Africa*, Cape Town: HSRC. Online. Available HTTP: <http://www.hsrcpress.ac.za/product.php?productid=2181> (accessed 10 July 2008).

Hall, Ruth; Jacobs, Peter and Lahiff, Edward. (2003). 'Evaluating land and agrarian reform in South Africa: Final Report', Occasional Paper, ELARSA no. 10: Cape Town: Programme for Land and Agrarian Studies, University of the Western Cape.

Hamadziripi, Alfred. (2007). SARPN (Southern Africa Regional Poverty Network) Alerts. E-mail *(*4 June 2007).

Hamilton, Sarah. (2002). 'Neoliberalism, gender and property rights in rural Mexico', *Latin American Research Review*: 37(1): 119–143.

Hammar, Amanda. (2003). 'The Making and unma(s)king of local government in Zimbabwe', in Amanda Hammar, Brian Raftopoulos and Stig Jensen (eds), *Zimbabwe's Unfinished Business: rethinking land, state and nation in the context of crisis*, Harare: Weaver Press: 119–54.

Hammar, Amanda, Raftopoulos, Brian and Jensen, Stig. (eds) (2003). *Zimbabwe's Unfinished Business*, London: Zed.

Hammel, Eugene. (1984). 'On the * * of studying household form and function', in R. Netting, R. Wilk and E. Arnould (eds) *Households: comparative and historical study of the domestic group*, Berkeley: University of California.

Handelman, Howard. (2009). *The Challenge of Third World Development*, 5th edn. Upper Saddle River, New Jersey: Pearson Prentice Hall.

Haney, Lynne. (1999). "But we are still mothers: gender, the state and the construction of need in postsocialist Hungary', in M. Burawoy and K. Verdery (eds) *Uncertain Transition: ethnographies of change in the postsocialist world*, Lanham, Md: Rowman and Littlefield: 151–97.

Hangar, Jean and Moris, Jon. (1973). *Mwea: an irrigated rice settlement in Kenya*, Munich: Weltforum.

Hanlon, Joseph. (1986). 'Producer cooperatives and the government in Zimbabwe', mimeo, London, unpublished.

———. (2002). *The Land Debate in Mozambique: will foreign investors, the urban elite, advanced peasants or family farmers drive rural development?* (cited in Palmer, 2005). Online. Available HTTP: <http://www.oxfam.org.uk/resources/learning/landrights/downloads/debatmoz.pdf > (accessed 30 January 2009).

———. (2004). 'Renewed land debate and the 'cargo cult' in Mozambique', *Journal of Southern African Studies*, September, 30(3): 603–25.

Hanstad, Tim, Nielsen, Robin and Brown, Jennifer. (2004). 'Land and Livelihoods: making land rights real for India's rural poor' Livelihoods Support Programme (LSP) Working Paper 12, Food and Agriculture Organisation/FAO of the United Nations: Geneva: FAO. Online. Available HTTP: <http://www.fao.org/sd/dim_pe4/pe4_040906_en.htm> (accessed 30 January 2009).

Harcourt, Wendy (2004) 'Women's networking for change: new regional and global configurations', *Journal of Interdisciplinary Gender Studies* 8(1–2): 120—32.

Harnecker, Marta. (2002). *Landless People—Building a Social Movement*. Online. Available HTTP: <http://www.rebelion.org/harnecker/landless300802.pdf> (accessed 16 December 2006).

Harris, Olivia. (1981). 'Households as natural units', in Kate Young, Judith Wolkowitz and Rosalyn McCullagh (eds) *Of Marriage and the Market*, London: Conference of Socialist Economists Books.

Harris, Olivia and Young, Kate. (1981). 'Engendered structures: some problems in the analysis of reproduction', in J. Kahn and J. Llobera (eds) *The Anthropology of Pre-Capitalist Societies*, London: Macmillan.

Harrison, Mark. (1982). 'Chayanov's theory of peasant economy', in John Harris (ed) *Rural Development*, London: Hutchinson.

Hartmann, Heidi. (1979). 'The unhappy marriage of Marxism and feminism: towards a more progressive union', *Capital and Class*: 8: 1–34.

Harvey, Neil. (1996). 'Impact of reforms to Article 27 on Chiapas: peasant resistance in the neoliberal public sphere' in Laura Randall (ed) *Reforming Mexico's Agrarian Reform*, Armonk, New York/London: M.E. Sharpe: 151–72.

Hatos, Adrian. (2006). Personal communication, Oradea, Romania.

Hayami, Y, Qisumbing, Agnes and Adriano, L. (1990). *Towards an Alternative Land reform Paradigm: a Philippines perspective*, Quezon City: Ateneo de Manila, University Presses.

Hellman, Judith. (1988). *Mexico in Crisis*, 2ⁿᵈ edn, New York: Holmes and Meier.

Hellum, Anne. (2007). *Human Rights, Plural Legalities and Gendered Realities: paths are made by walking*, Harare: University of Zimbabwe Press.

Hellum, Ann and Derman, Bill. (2004). 'Land reform and human rights in contemporary Zimbabwe: balancing individual and social justice through an integrated human rights framework', *World Development*, 32(10): 1785–1805.

Hendricks, Fred. (2003). 'Land inequality in democratic South Africa', in M. M. Bell and F. Henricks with A. Bacal (eds) *Walking Towards Justice: democratization in rural life*, Oxford and Amsterdam: Elsevier.

Henson, Henny. (1984). Personal communication, Copenhagen; citing MA thesis on Gender and Land Resettlement, University of Copenhagen, Denmark.

Herbst, Jeffrey. (1991). 'The dilemmas of land policy in Zimbabwe', *Africa Insight*, 21(4): 269–76.

Herring, Ronald. (1990). 'Explaining anomalies in agrarian reform: lessons from South India' in Roy Prosterman, Mary Temple and Timothy Hanstad (eds)

Agrarian Reform and Grassroots Development: ten case studies, Boulder: Lynne Rienner.

Hershatter, Gail. (2000). 'Local meanings of gender and work in rural Shaanxi in the 1950s', in B. Entwisle and G. Henderson (eds) *Re-Drawing Boundaries: work, household and gender in China*, Berkeley: University of California Press.

———. (2002). 'The gender of memory: rural Chinese women and the 1950s', *Signs*: 28(1): 43–69.

Hilhorst, Thea. (2001). 'Women's land rights: current developments in sub-Saharan Africa', in C. Toulmin and J. Quan (eds) *Evolving Land Rights*, London: IIED.

Himmelweit, Susan and Mohun, Simon. (1977). 'Domestic labour and capital' *Cambridge Journal of Economics*: 1(1): 15–31.

Hindess, Barry and Hirst, Paul. (1975). *Pre-Capitalist Modes of Production*, London: Routledge and Kegan Paul.

———. (1977). *Modes of Production and Social Formation*, London: Macmillan.

Hinton, William. (1966). *Fanshen*, London: Penguin.

———. (1998). 'The importance of land reform in the reconstruction of China', *Monthly Review*, 50(3): 147–60.

Hirschmann, David. (1998). 'Civil society in South Africa: learning from gender themes', *World Development* 26(2): 227–38.

Hirschon, Renée. (1984). 'Introduction: property, power and gender relations', in R. Hirschon (ed) *Women and Property—Women as Property*, Beckenham, Kent: Croom Helm.

Hobsbawm, Eric (ed) (1980) *Peasants in History: essays in honour of Daniel Thorner*, Oxford: Oxford University Press.

Hobsbawm, Eric and Ranger, Terence (eds) (1983) *The Invention of Tradition*, Cambridge: Cambridge University Press.

Holmes, Rebecca and Slater, Rachel. (2008). 'Measuring progress on gender and agriculture in the 1982 and 2008 World Development Reports', *Gender and Development*, 16(1): 27–40.

Holzner, Brigitte. (1995). 'Research on gender issues and agrarian change in East-Central and Eastern European countries: some preliminary impressions', *Vena*, 7(1): 12–17.

———. (2008). 'Agrarian restructuring and gender—designing family farms in Central and Eastern Europe', *Gender, Place and Culture* 15(4): 431–43.

Hong, Zhang. (2008). 'Back to the land—again', *The Guardian*, 27 October. Online. Available HTTP: <http://www.guardian.co.uk/commentisfree/2008/oct/27/china-economy> (accessed 19 December 2008).

Honig, Emily. (2000). 'Iron girls revisited' in B. Entwisle and G. Henderson (eds) *Re-Drawing Boundaries: work, household and gender in China*, Berkeley: University of California Press: 97–110.

hooks, bell. (1982). *Ain't I a Woman*, London: Pluto.

Hooper, J. (1996a). 'Women left chained to agriculture and poverty', *The Guardian*, 12 November.

Hooper, J. (1996b). 'Free market summit condemned', *The Guardian*, 18 November.

Houtart, F. and Lemercinier, G. (1984). *Hai Van: life in a Vietnamese commune*, London: Zed.

Howden, Daniel. (2006). 'Dead by 34: how Aids and starvation condemn Zimbabwe's women to early grave', *The Independent*, 17 November 2006. Online. Available HTTP: <http://news.independent.co.uk/world/africa/article/1990401.ece> (accessed 12 September 2007).

Howell, Jude. (2000). 'Shifting relationships and competing discourses in Post-Mao China: the All-China Women's Federation and the People's Republic' in

S. Jacobs, R. Jacobson and J. Marchbank (eds) *States of Conflict: gender, violence and resistance,* London: Zed: 125–43.

——. (2002). 'Women's political participation in China: struggling to hold up half the sky', *Parliamentary Affairs:* 55: 43–56.

——. (2006). 'Reflections on the Chinese state', *Development and Change,* 37(2): 273–97.

Huber, Sophie (1990/91) 'Peruvian land reform and rural women', *J. of Legal Pluralism and Unofficial Law,* 30/31, Special Double issue: 165–222.

Huizer, Gerrit. (2001). 'Peasant Mobilization for land reform: historical cases and theoretical considerations', in K. Ghimire (ed) *Land Reform and Peasant Livelihoods,* London: ITDG/Intermediate Technology Development Group.

Human Rights Watch. (2008). 'Zimbabwe: surge in state-sponsored violence', April 2008. Online. Available HTTP: <http://www.hrw.org/en/news/2008/04/24/ zimbabwe-surge-state-sponsored-violence> (accessed 30 January 2009).

Hyden, Göran. (1986). 'The invisible economy of smallholder agriculture in Africa' in J. Moock (ed) *Understanding Africa's Rural Households and Farming Systems,*: 11–35. Boulder: Westview

IFAD (International Fund for Agricultural Development). (2001). *Rural Poverty Report,* Oxford: Oxford University Press.

IFPRI. (2002). 'The high price of gender inequality', *IFPRI Perspectives,* 24, Online. Available HTTP: <http://www.ifpri.org/REPORTS/02spring/02springc.htm> (accessed 30 January 2009).

Ikdahl, Ingunn; Hellum, Anne; Kaarhus, Randi; Benjaminsen, Tor A. and Kameri-Mbote, Patricia. (2006). Human rights, formalisation and women's land rights in southern and eastern Africa, *Studies in Women's Law* no. 57, University of Oslo. Online. Available HTTP: <http://www.ielrc.org/content/w0507.pdf> (accessed 30 January 2009).

ILO. (1999). 'Report of the Director-General: Decent Work, Geneva: ILO. Online. Available: HTTP <http://www.ilo.org/public/english/standards/relm/ilc/ilc87/ rep-i.htm> (accessed 2 February 2009).

Independent Review of Land Issues [Shaun Williams]. (2005). Vol. II, no. I (Southern Africa), December. Online. Available HTTP: <http://www.oxfam.org.uk/ what_we_do/issues/livelihoods/landrights/downloads/ind_land_newsletter_ sth_afr_june_2004.rtf> (accessed 30 January 2009).

Inhetveen, H. and Blasche, M. (1988). 'Women in the smallholder economy', in T. Shanin (ed) *Peasants and Peasant Societies,* Harmondsworth: Penguin, 2ⁿᵈ ed: 28–34

International Crisis Group. (2005). Zimbabwe's Operation *Murambatsvina:* the tipping point?, Report no. 97, 17 August, Brussels: ICG. Online. Available HTTP: <http://www.crisisgroup.org/home/index.cfm?id=3618> (accessed 30 January 2009).

Ireson, Carol. (1996). *Field, Forest and Family: women's work and power in rural Laos,* Boulder: Westview.

IRIN News. (2003). 'Zimbabwe: Land reform beneficiaries under scrutiny' 24 February. Online. Available HTTP: <www.irinnews.org/report. asp?ReportID=32476> (accessed 29 March, 2006).

——. (2007a). 'Rural living standards now apply in the capital' 2 August. Online. Available HTTP: <http://www.irinnews.org/Report.aspx?ReportId=73551> (accessed 26 August 2008)

——. (2007b). 'Overzealous police impound food aid' 3 August. Online. Available HTTP: <http://www.irinnews.org/report.aspx?ReportID=73578 > (accessed 26 August 2008).

——. (2008a). 'Zimbabwe: New land owners face eviction' 11 February. Online. Aavailable HTTP: <www.irinnews.org/report.aspx?reportid=76682> (accessed 18 July 2008).

————. (2008b). 'Zimbabwe: Small scale farmers seen as backbone of food security' 15 May. Online. Available HTTP: <www.irinnews.org/report. aspx?reportid=78222> (accessed 19 July 2008)

Izumi, Kaori. (1999). 'Liberalisation, gender, and the land question in sub-Saharan Africa', *Gender and Development* 7(3): 9–18.

————. (2006). *The Land and Property Rights of Women and orphans in the Context of HIV and AIDS: case studies from Zimbabwe*, Pretoria/Tshwane: Human Sciences Research Council. Online. Available HTTP: <http://www. hsrcpress.ac.za/product.php?productid=2167> (accessed 2 February 2009).

Jacka, Tamara. (1997). *Women's Work in Rural China*, Cambridge: Cambridge University Press.

Jackson, Cecile. (1996). Rescuing gender from the poverty trap. *World Development* 24(3): 489–504.

Jacobs, Susie. (1983). 'Women and land resettlement in Zimbabwe', *Review of African Political Economy*, 27/28: 33–50.

————. (1989). *Gender Relations and Land Resettlement in Zimbabwe*, D. Phil thesis., Institute of Development Studies, University of Sussex.

————. (1992). 'Gender and land reform: Zimbabwe and some comparisons', *International Sociology*, 7(1): 5–34.

————. (1995). 'Changing gender relations in Zimbabwe: experiences of individual family resettlement schemes', in Diane Elson (ed) *Male Bias in the Development Process*, Manchester: Manchester University Press, 2nd ed.

————. (1996). ' 'The gendered politics of land reform: three comparative studies', in Vicky Randall and Georgina Waylen *Gender, Politics and the State*, London: Routledge, 121–42.

————. (1997). 'Land to the tiller? gender relations and land reforms', *Society in Transition* (formerly, *J. of S. African Sociology*) 1(1–4): 82–100.

————. (2000a). 'Globalisation, states and women's agency: possibilities and pitfalls', in S. Jacobs, R. Jacobson and J. Marchbank (eds) *States of Conflict: gender, violence and resistance*, London: Zed.

————. (2000b). 'The Effects of Land Reform on Gender Relations in Zimbabwe'. In *Land Reform in Zimbabwe: Constraints and Prospects*, eds. T.A.S. Bowyer-Bower and C. Stoneman. Aldershot: Ashgate.

————. (2002). 'Land Reform: still a goal worth pursuing for rural women?, *Journal of International Development*, 14: 887–98.

————. (2003). 'Democracy, class and gender in land reforms: a Zimbabwean example', in M. Bell and F. Hendricks (eds) *Walking Towards Justice: democratization in rural life*, Oxford and Amsterdam: JAI/Elsevier: 203–29.

————. (2004a). 'New forms, longstanding issues and some successes', *J. of Interdisciplinary Gender Studies, Special Issue: Feminist Organisations and Networks in a Globalising Era*, 8(1–2): 171–94.

————. (2004b). 'Livelihoods, Security and Needs: Gender and Land Reform in South Africa' *J. International Women's Studies*, November, 6(1): 1–19. Online. Available HTTP: www.bridgew.edu/SoAS/jiws/nov04/genderandlandSouthAfrica (accessed 29 June 2009).

————. (2008). 'Women behaving badly: still a taboo subject? Women's roles in ethno-political violence', paper presented at ESRC seminar series Women and Violent Conflict: Intersections of 'Race', Ethnicity and Gender, Coventry, University of Warwick, 2–3 May.

Jacobs, Susie and Howard, Tracy. (1987). 'Women in Zimbabwe: stated policy and state action', in Haleh Afshar (ed) *Women, State and Ideology: studies from Africa and Asia*, New York: University of New York Press: 28–47.

Jacobs, Susie; Jacobson, Ruth and Marchbank, Jennifer (eds) (2000) *States of Conflict: gender, violence and resistance*, London: Zed Books.

Jacobson, Ruth. (1999). 'Integrating gender into the anlaysis of the Mozambiquan conflict', *Third World Quarterly,* Special Issue, *Complex Political Emergencies:* 20(1): 175–88.

James, Deborah. (2007). "Failure' of land reform should be seen in context', *Mail and Guardian online,* 11/9/07. Online. Available HTTP: <http://www.mg.co.za/article/2007–09–11-failure-of-land-reform-should-be-seen-in-context> (accessed 27 July 2008).

de Janvry, Alain; Gordillo, Gustovo and Sadoulet, Elisabeth (eds). (1997). *Mexico's Second Agrarian Reform: household and community responses: 1990–1994,* San Diego: Ejido Reform Research Project, University of California at San Diego.

de Janvry, Alain; Platteau, Jean-Phillippe; Gordillo, Gustavo and Sadoulet, Elisabeth. (2001). 'Access to land and land policy reforms' in A. de Janvry, G. Gordillo, J. P. Platteau and E. Sadoulet (eds) *Access to Land, Rural Poverty and Public Action,* Oxford: Oxford University Press.

de Janvry, Alain; Sadoulet, Elisabeth; David, Benjamin and Gordillo de Anda, Gustavo. (1996). 'Ejido sector reforms: from land reform to rural development', in L. Randall (ed) *Reforming Mexico's Agrarian Reform,* Armonk, New York/London: M.E.Sharpe: 71–106.

Jaquette, Jane S., and Summerfield, Gale (eds) (2006). *Women and Gender Equity in Development Theory and Practice: institutions, resources, and mobilization,* Durham, N.C.: Duke University Press.

Jashok, Maria and Miers, Suzanne. (1994). 'Traditionalism, continuity and change', in M. Jaschok and S. Miers (eds) *Women and Chinese Patriarchy,* London: Zed.

Jassal, Smita. (2004). 'Limits of empowerment: Mallahin Fishponds in Madhubani, Bihar', in S. Krishna (ed) *Livelihood and Gender: equity in community resource management,* New Delhi: Sage.

Jayawardena, Kumari. (1986). *Feminism and Nationalism in the Third World,* London: Zed Books.

Jenkins, Rhys. (2006). 'Globalization, FDI and employment in Viet Nam', *Transnational Corporations,* 15(1): 115–42.

Johnson, Kay Ann. (1983). *Women, the Family and Peasant Revolution in China,* Chicago: University of Chicago Press.

Johnson, R.W. (2007). 'In time of famine', *London Review of Books,* 22/2/07.

Judd, Ellen. (1994). *Women's Work in Rural North China,* Stanford: Stanford University Press.

Kabeer, Naila. (1994). *Reversed Realities: gender hierarchies in development theory,* London: Verso.

Kabeer, Naila and Tran Thi van Anh. (2002). 'Leaving the rice fields, but not the countryside', in S. Razavi (ed) *Shifting Burdens: gender and agrarian change under neoliberalism,* Bloomfield, Connecticut: Kumarian Press: 109–50.

Kagwanja, Peter. (2007). 'Land reform: the art of the possible', *Mail & Guardian Online,* 6/7/2008. Online. Available HTTP: <http://www.mg.co.za/article/2007–04–19-land-reform-the-art-of-possible> (accessed 2 February 2009).

Kahn, Joel. (1975). 'Economic scale and the cycle of petty commodity production in West Sumatra', in M. Bloch (ed) *Marxist Analyses and Social Anthropology,* London: Malaby Press.

———. (1978). 'Ideology and social structure in Indonesia', *Comparative Studies in Society and History,* 20(1): 103–22.

———. (1980). *Minangkabau Social Formations: Indonesian peasants and the world-economy,* Cambridge: Cambridge University Press.

———. (1982). 'From peasants to petty commodity production in Southeast Asia', *Bulletin of Concerned Asian Scholars,* (14)1: 13–15.

Kalugina, Zamfira. (2000). 'Paradoxes of agrarian reform in Russia', in L. A. Norsworthy (ed) *Russian Views of Transition in the Rural Sector,* Washington, DC: World Bank.

Kampwirth, Karen. (1998). 'Legislating personal politics in Sandinista Nicaragua', *Women's Studies International Forum,* 21(1): 53–64.

Kandiyoti, Deniz. (1988). 'Bargaining with patriarchy', *Gender and Society,* 11(3): 274–90.

——. (1989). 'Women and household production: the impact of rural transformation in Turkey', in K. and P. Glavanis (eds) *The Rural Household in the Middle East,* London: Zed.

——. (1998). 'Rethinking bargaining with patriarchy', in C. Jackson and R. Pearson (eds) *Feminist Visions of Development,* London: Routledge.

——. (2003). 'The cry for land: agrarian reform, gender and land rights in Uzbekistan' *J. Agrarian Change,* 3(1–2): 225–56.

Katerere, Fred. (2008). 'Zimbabwean women sell sex for food in Mozambique', *Mail & Guardian Online,* 23/5/08. Online. Available HTTP: <http://www.mg.co.za/article/2008-05-23-zimbabwean-women-sell-sex-for-food-in-Mozambique> (accessed 4 August 2008).

Kautsky, Karl (1988 [1899]) *The Agrarian Question,* London: Unwin Hyman.

Kawamara-Mishambi, Sheila and Ovonji-Odida, Irene. (2003). 'The 'lost clause': the campaign to advance women's property rights in the Uganda Land Act (1998)', in A. M. Goetz and S. Hassim (eds) *No Shortcuts to Power: African women in politics and policy-making,* London/Cape Town: Zed and David Philip.

Kay, Cristóbal. (1999). 'Rural development: from agrarian reform to neoliberalism and beyond', in R. Gwynne and C. Kay (eds) *Latin America Transformed,* London: Edward Arnold.

——. (2001). 'Reflections on rural violence in Latin America', *Third World Quarterly,* 22(5): 741–75.

——. (2002). 'Chile's neoliberal agrarian transformation and the peasantry', *Journal of Agrarian Change,* 2(4): 464–501.

Kazembe, Joyce. (1986). 'The woman issue", in Ibbo Mandaza (ed) *Zimbabwe: the political economy of transition,* Dakar: Codeseria.

Kearney, Michael. (1996). *Reconceptualising the Peasantry: anthropology in global perspective,* Boulder: Westview.

Kelkar, Govind and Gala, Chetna. (1990). 'The Bodhgaya land struggle', in A. Sen (ed) *A Space in the Struggle,* Delhi: Kali for Women.

Kerkvliet, Benedict J. Tria. (2006). 'Agricultural land in Vietnam: markets tempered by family, community and socialist practices', *J. Agrarian Change,* 6(3): 285–305.

Kerkvliet, Benedict J. Tria and Porter, D.J. (eds.) (1995) *Viet Nam's rural transformation,* Boulder: Westview Press.

Kidder, Thea. (1997). 'Micro-finance and women: how to improve food security?', *Links,* Oxfam/UKI Newsletter, October: 5.

King, R. (1977). *Land Reforms: a world survey,* London: Bell and Son.

Kinsey, Bill. (1982). 'Forever gained: resettlement and land policy in the context of national development in Zimbabwe', *Africa,* 52(3): 92–113.

——. (1999). 'Land reform, growth and equity: emerging evidence from Zimbabwe's resettlement programme', *Journal of Southern African Studies,* 25(2): 173–96.

——. (2000). 'The implications of land reform for rural welfare' in T.A.S. Bowyer-Bower and Colin Stoneman (eds) *Land Reform in Zimbabwe: Constraints and prospects,* Aldershot: Ashgate.

Kinsey, Bill. (2001). Land invasions Britain-Zimbabwe Society Research Day, June, Oxford unpublished.

——. (2004a). 'Guest editor's introduction', *World Development,* 32(10): 1663–1667.

——. (2004b). 'Zimbabwe's land reform program: underinvestment in post-conflict transformation', *World Development,* 32(10): 1669–1696.

——. (2006). 'Land issues 1. Fast track down the furrow', *Zimbabwe Society Newsletter,* 6(1): 1–3.

Kodoth, Praveena. (2004). 'Gender aspects of family property and land rights: regulation and reform in Kerala', in S. Krishna (ed) *Livelihood and Gender: equity in community resource management,* New Delhi: Sage.

——. (2005). 'Fostering insecure livelihoods: dowry and female seclusion in left developmental contexts in West Bengal and Kerala', *Economic and Political Weekly,* 18 June: 2543–2554.

Koopman, Jeanne. (1997). 'The hidden roots of the African food problem: looking within the rural household', in N. Visvanathan, L. Duggan, L. Nisonoff, and N. Wiegersma (eds) *The Women, Gender and Development Reader,* London: Zed Books: 132–41.

Korinek, Kim. (2003). *Women's and Men's Economic Roles in Northern Viet Nam during a Market Reform,* PhD thesis, Seattle: University of Washington.

Kubatana.net. (2008). 'Operation glossary—a guide to Zimbabwe's internal campaigns', *IRIN News,* 1 May. Online. Available HTTP: <http://www.kubatana.net/html/archive/demgg/080501irin.asp?sector=DEMGG> (accessed 10 July, 2008).

Kusterer, Ken. (1990). 'The imminent demise of patriarchy', in I. Tinker (ed) *Persistent Inequalities,* Oxford: Oxford University Press: 239–56.

Kwidini, Tonderai. (2008). 'Zimbabwe's tobacco sector fizzles out', *Mail & Guardian Online,* 2/7/08. Online. Available HTTP: <http://www.mg.co.za/article/2008–07–02-zimbabwes-tobacco-sector-fizzles-out> (4 August 2008).

LaClau, Ernesto. (1971). 'Feudalism and capitalism in Latin America', *New Left Review,* 67: 19–38.

Lago, María Soledad. (1987). 'Rural women and the neo-liberal model in Chile', in C. D. Deere and M. León de Leal (eds) *Rural Women and State Policy,* Boulder: Westview: 21–33.

Lahiff, Edward and Cousins, Ben. (2005). 'Smallholder agriculture and land reform in South Africa', *IDS Bulletin,* 36(2): 127–31.

Lapp, Nancy. (2004). *Landing Votes: representation and land reform in Latin America,* New York and Houndmills: Palgrave Macmillan.

Laslett, Peter. (1984). 'The family as a knot of individual interests', in R. M. Netting, R. Wilk and E. Arnould (eds) *Households: comparative and historical study of the domestic group,* Berkeley: University of California Press.

Lastarría-Cornhiel, Susana. (1997). 'Impact of privatization on gender and property rights in Africa', *World Development* 25(8): 1317–1333.

Lastarría-Cornhiel, Susana and García-Frias, Zoraida (2005) 'Gender and Land Rights: Findings and Lessons from Country Studies' in FAO Gender and Land Compendium of Country Studies, Rome: FAO of the UN.

Le Thi Nham Tuyet. (1996). 'Rural women and the Red River Delta: gender, water management and economic transformation', Hanoi: Unpublished document.

Leach, Melissa. (2007). 'Earth mother myths and other ecofeminist fables: how a strategic notion rose and fell', *Development and Change* 38(1): 67–85.

Lebert, Tom. (2006). 'An introduction to land and agrarian reform in Zimbabwe', in Peter Rosset, Raj Patel and Michael Courville (eds) *Promised Land: competing visions of agrarian reform,* Oakland: Food First Books: 40–56.

Lee, Chin Kwan. (2005). 'Livelihood Struggles and Market Reform: (un)making Chinese labour after state socialism', UNRISD Occasional paper 2: Geneva, UNRISD. Online. Available HTTP: http://www.unrisd.org/80256B3C005BCCF9/(httpPublications)/755EB01A0C1A165BC125700E00380454?OpenDocument> (accessed 2 February 2009).

Lee-Smith and Trujillo, Cataline Hinchey. (2006). 'Unequal rights: women and property', in J. S. Jaquette, and G. Summerfield (eds) *Women and Gender Equity in Development Theory and Practice: institutions, resources, and mobilization*, Durham, North Carolina: Duke University Press: 159–72.

Lehmann, David. (1974). *Agrarian Reform and Agrarian Reformism*, London: Faber and Faber.

Lehmann, David. (1986). 'Two paths of agrarian capitalism, or a critique of Chayanovian Marxism, *Comparative Studies in Society and History*, 28(4): 601–27.

Lenin, V. I. (1977). *The Development of Capitalism in Russia*, Moscow: Progress Publishers.

Levins, Richard. (2005). 'How Cuba is going ecological', *Capitalism, Nature, Socialism*, 16(3): 7–25.

Lewin, Moshe. (1968). *Russian Peasants and Soviet Power*, London: Allen and Unwin.

Li Huayin. (2005). 'Life cycle, labour remuneration and gender inequality in a Chinese agrarian collective', *J. Peasant Studies*, 32(2): 277–303.

Li Jianghong and Lavely, William. (2003). 'Village context, women's status and son preference among rural Chinese women', *Rural Sociology* 68(1): 87–106.

Li Weisha. (1999). 'Changes in housing patterns for rural Chinese women' in I. Tinker and G. Summerfield (eds) *Women's Rights to House and Land: China, Laos, Vietnam*, Boulder: Lynne Rienner: 231–40.

Li Zongmin. (1999). 'Changing land and housing use by rural women in northern China', in I. Tinker and G. Summerfield (eds) *Women's Rights to House and Land: China, Laos, Vietnam*, Boulder: Lynne Rienner: 241–64.

Li Zongmin and Bruce, John. (2005). 'Gender, landlessness and equity in China' in P. Ho (ed) *Developmental Dilemmas: Land reform and institutional change in China*, Abingdon, Oxon: Routledge

Liljeström, Rita; Lindskog, Eva; Ang, Nguyen van and Tinh, Vuong Xuan. (1998). *Profit and Poverty in Rural Vietnam: winners and losers of a dismantled revolution*, Richmond, Surrey: Nordic Institute of Asian Studies and Curzon.

Lin, Justin. (1999). 'China: farming institutions and rural development', in M. Meurs (ed) *Many Shades of Red*, Lanham, Md: Rowman and Littlefield:

Lin, Qi. (2004). 'Rural women venture into politics in China', *China Daily*, 27 December. Online. Available HTTP: http://www.chinadaily.com.cn/english/doc/2004–12/27/content_403517.htm> (accessed 2 February 2009).

Lindio-McGovern, Ligagya. (1997). *Filipino Peasant Women: exploitation and resistance*, Philadephia: University of Pennsylvania Press.

Linkogle, Stephanie. (2001). 'Nicaraguan women in the age of globalization', in S. Rowbotham and S. Linkogle (eds) *Women Resist Globalization*, London: Zed.

Lipton, Michael. (1974). 'Towards a theory of land reform', in D. Lehmann (ed), *Peasants, Landlords and Governments: agrarian reform in the third world*, New York: Holmes and Meier.

Lipton, Michael. (1993). 'Land reform as commenced business: the evidence against stopping', *World Development*, 21(4): 641–57.

Lipton, Michael. (2005). 'The family farm in a globalizing world: the role of crop science in alleviating poverty', in IFPRI (International Food Policy Research Institue): 2020 Vision discussion paper 40. Online. Available HTTP: <http://www.ifpri.org/2020/dp/vp40.asp> (accessed 2 February 2009).

Liu, Xin. (2000). *In One's Own Shadow: an ethnographic account of the condition post-reform China*. Berkely University of California Press.

Llambí, Louis. (1988). 'Small modern farmers: neither peasants nor fully-fledged capitalists?', *J. Peasant Studies*, 15(3): 350–72.

Logo, Patrice and Bikie, Elise-Henriette. (2003). 'Women and land in Cameroon', in L. Muthoni Wanyeki (ed) *Women and Land in Africa*, London: Zed.

Long, Norman (ed) (1984) *Family and Work in Rural Societies*, London: Tavistock.

———. (1992). 'From paradise lost to paradise regained? The case for an actor-oriented sociology of development', in N. Long and A. Long (eds) *Battlefields of Knowledge: the interlocking of theory and practice in social research and development*, London: Routledge.

Low, A. (1986). *Agricultural Development in Southern Africa: farm household theory and the food crisis*, London: James Currey.

Lund, R. (1978). *Prosperity to Mahaweli: a survey on women's working and living conditions in a settlement area*, Colombo, Sri Lanka: People's Bank Research Division.

Luong, Hy Van. (1989). 'Vietnamese kinship: structural principles and the socialist transformation in Northern Vietnam', *Journal of Asian Studies*, 48(4): 741–56.

———. (2003a). 'Introduction: postwar Vietnamese society', in H. Van Luong (ed) *Postwar Vietnam: dynamics of a transforming society*, Lanham, Md./Singapore: Rowman and Littlefield/Institute of Southeast Asian Studies: 1–26

———. (2003b). 'Gender relations: ideologies, kinship practices and political economy', in H. Van Luong (ed) *Postwar Vietnam: dynamics of a transforming society*, Lanham, Md./Singapore: Rowman and Littlefield/Institute of Southeast Asian Studies: 201–24.

Luong, Hy Van and Unger, Jonathan. (1998). 'Wealth, power and poverty in the transition to market economies: the process of socio-economic differentiation in rural China and Northern Vietnam', *The China Journal*, 40, July: 61–93.

Luxemburg, Rosa. ([1913] 1973). *The Accumulation of Capital: an anti-critique*, New York: Monthly Review Press.

Luz, M. Padilla, Murguíalday, C. and Crillon, A. (1987). 'Impact of the Sandinista reform on rural women's subordination', in C. D. Deere, and M. León de Leal (eds) *Rural Women and State Policy*, Boulder: Westview: 124–41.

McCall, Michael. (1987). 'Carrying heavier burdens but carrying less weight: the implications of villagisation for women in Tanzania', in J. Momsen and J. Townsend (eds) *The Geography of Gender in the Third World*, London: Hutchinson: 192–214.

McCusker, Brent. (2004). 'Land use and cover change as an indicator of transformation on recently redistributed farms in Limpopo Province, South Africa', *Human Ecology*, 32(1): 49–75.

McFadden, Patricia. (2005). 'Becoming postcolonial: African women changing the meaning of citizenship', *Meridians: Feminism, Race, Transnationalism*, 6(1): 1–22.

MacGarry, Brian. (2007). 'The Zimbabwe economy in 2006', *Zimbabwe Review*, (2): 1–23 May.

———. (2008). 'The Zimbabwe economy in 2007', *Zimbabwe Review*, (2): 1–24 May.

McGreal, Chris. (2008a). 'Zimbabwe's voters told: choose Mugabe or you face a bullet', *The Guardian*, 18 June: 1–2.

———. (2008b). 'What comes after a trillion?' *The Guardian* G2,12–15, 18 July.

———. (2008c). 'Threat of mass starvation looms in Zimbabwe after latest harvest fails', *The Guardian*, 21/7/08. Online. Available HTTP: <http://www.guardian.co.uk/world/2008/jul/21/zimbabwe.unitednations> (accessed 2 February 2009).

———. (2009). 'How did it come to this?' *Guardian* G2, 13 February: 10–13.

McGreal, Chris and McVeigh, Tracy. (2007). 'Mugabe allies 'set up' political terror', *The Guardian*, 22/06/08. Online. Available HTTP: <http://www.guardian.co.uk/world/2008/jun/22/zimbabwe> (accessed 4 August 2008).

Mackintosh, Maureen. (1981). 'Gender and economics: the sexual division of labour and the subordination of women', in K. Young, A. Wolkowitz and R. McCullagh (eds) *Of Marriage and the Market*, London: CSE Books: 1–15.

Magardie, Khadija. (1999). 'Women—still the second sex', *Mail and Guardian*, 29 December.

Mail and Guardian. (2006). 'DA warns against changes to land-reform policy', *Mail and Guardian Online*, 16/10/2006. Online. Available HTTP: <http://www.mg.co.za/article/2006–10–16-da-warns-against-changes-to-landreform-policy> (accessed 2 February 2009).

Maine, Henry. ([1861]) 2001). *Ancient Law*. N. Jersey: Transaction Publishers.

Mail and Guardian. (2007). 'Slow land reform leads to grabs', *Mail and Guardian Online*, 5/7/07. Online. Available HTTP: <http://www.mg.co.za/article/2007–07–05-slow-land-reform-leads-to-grabs> (accessed 6 July 2008).

Malarney, Shaun Kingsley. (2003). 'Return to the past? the dynamics of contemporary religious and ritual transformation', in H. Van Luong (ed) *Postwar Vietnam: dynamics of a transforming society*, Lanham, Md./Singapore: Rowman and Littlefield/Institute of Southeast Asian Studies: 225–56.

Manimala. (1983). 'Zameen Kenkar? Jote Onkar!' The story of women's participation in the Bodhaya struggle', *Manushi*, 14(1): 2–16.

Manji, Ambreena (2003a) 'Remortgaging women's lives: the World Bank's land agenda in Africa', *Feminist Legal Studies* 11: 139–63.

———. (2003b). 'Capital, labour and land relations in Africa: a gender analysis of the World Bank's Policy Research Report on Land Institutions and Land Policy', *Third World Quarterly*, 24(1): 97–114.

———. (2006). *The Politics of Land Reform in Africa: from communal tenure to free markets*, London: Zed Books.

Mann, Michelle. (2000). 'Women's access to land in the former Bantustans: constitutional conflict, customary law, democratisation and the role of the state', Land Reform and Agrarian Change in southern Africa, Occasional Paper No. 15, PLAAS (Programme for Land and Agrarian Studies), Cape Town: University of the Western Cape.

Mao Tse-tung. (1960). *Report from Xunwu*. (in Chinese: cited in Bossen, 2002).

Marate, Kanta. (2004). 'People's land reform initiatives: a note on the Budelkhand-Baghelkhand Area, Madhya Pradesh' in S. Krishna (ed) *Livelihood and Gender: equity in community resource management*, New Delhi: Sage.

Marcus, Tessa; Eales, Kath and Wildschut, Adele. (1996). *Down to Earth: land demand in the new South Africa*, Durban: Indicator Press.

Martin, D and Beitel, M. (1987). 'The hidden abode of reproduction: conceptualisation of the household in Southern Africa', *Development and Change*, 18(2): 215–34.

Martin, JoAnn. (1994). 'Antagonisms of class and gender in Morelos', in H. Fowler-Salmini and M. K. Vaughan (eds) *Women of the Mexican Countryside: 1850–1990*, Tuscon: University of Arizona Press.

Martinez, Philip. (1993). 'Peasant policy within the Nicaraguan agrarian reform', *World Development*, 21(3): 475–87.

Martins, José de Souza. (2003). 'Representing the peasantry? struggles for/about land in Brazil', *J. of Peasant Studies*, 29(3–4): 300–35.

Marx, Karl. (1967). *The Eighteenth Brumaire of Louis Bonaparte*, Moscow: Progress Publishers.

Masiiwa, Medicine. (2004). *Land Reform Programme in Zimbabwe: disparity between policy design and implementation*, Harare: Institute of Development Studies.

May, Julian. (2000). *Poverty and Inequality in South Africa: meeting the challenge*, New York: Zed Books.

May, Julian; Rogerson, Chris and Vaughan, Megan. (2000). 'Livelihoods and assets', in J. May (ed) *Poverty and Inequality in South Africa*, Cape Town: D. Philip.

May, Julian; Roberts, Benjamin; Govender, Juby and Gayadeen, Priya. (2000). 'Monitoring and Evaluating the Quality of Life of Land Reform Beneficiaries: 1998–99', Technical Report prepared for the Department of Land Affairs, Pretoria, S. Africa

Mayoux, Linda. (1993). 'Integration is not enough: gender inequality and empowerment in Nicaraguan agricultural cooperatives', *Development Policy Review*, 11 (1): 67–89.

Meillasoux, Claude. (1972). 'From reproduction to production', *Economy and Society*, 1 (2): 93–105.

———. (1981). *Maidens, Meal and Money: capitalism and the domestic economy*, Cambridge: Cambridge University Press.

Meldrum, Andrew. (2003). 'Mugabe's cronies 'grab land'', *The Guardian*, 22/2/03. Online. Available HTTP: <http://www.guardian.co.uk/world/2003/feb/22/zimbabwe.andrewmeldrum> (accessed 7 August 2008).

———. (2007). 'US predicts regime change in Zimbabwe as hyperinflation destroys the economy', *The Guardian Online*, 22/6. Online. Available HTTP: <http://www.guardian.co.uk/money/2007/jun/22/Zimbabwenews> (accessed 4 September 2008).

———. (2008). 'Mutiny in the offing?', *The Guardian Online*, 31/3, Online. Available HTTP: <http://www.guardian.co.uk/commentisfree/2008/mar/31/mutinyintheoffing> (accessed 3 September 2008).

Metoyer, Cynthia Chávez. (2000). *Women and the State in Post-Sandinista Nicaragua*, Boulder: Lynne Rienner.

Meurs, Mieke, (1992) 'Popular participation and central planning in Cuban socialism: the experience of agriculture in the 1980s', *World Development*, 20(2): 229–40.

———. (1997). 'Downwardly mobile: women in the decollectivisation of East European agriculture', in N. Visvanathan, L. Duggan, L. Nisonoff and N. Wiegersma (eds) *The Women, Gender and Development Reader*, London: Zed Books.

———. (1999). 'The continuing importance of collective agriculture', in M. Meurs (ed) *Many Shades of Red: state policy and collective agriculture*, Lanham, Md: Rowman and Littlefield.

———. (1999). 'Conclusion: looking forward', in M. Meurs (ed) *Many Shades of Red: state policy and collective agriculture*, Lanham, Md: Rowman and Littlefield.

Meurs, Mieke; Koizhouharova, Veska and Stoyanova, Rositsa. (1999). 'Bulgaria: from cooperative village to agro-industrial complex', in M. Meurs (ed) *Many Shades of Red: state policy and collective agriculture*, Lanham, Md: Rowman and Littlefield 87–124.

Mgugu, Abby (Women's Land Lobby, Zimbabwe). (2002). Interview, Johannesburg, November.

Michelson, Ethan and Parish, William. (2000). 'Gender differentials in economic success: rural China in 1991', in B. Entwisle and G. Henderson (eds) *Re-Drawing Boundaries: work, household and gender in China*, Berkeley: University of California Press: 134–56.

Mies, Maria. (1986). *Patriarchy and Accumulation on a World Scale*, London: Zed Books.

Millett, Kate. (1970). *Sexual Politics*, Garden City, New York: Doubleday.

Mintz, Sidney. (1973). 'A note on the definition of peasantries', *J. Peasant Studies*, 1(1): 91–106.

———. (1982). 'Descrying the peasantry', *Review*, VI (2): 209–25.

Mirza, Heidi S. (ed) (1997) *Black British Feminism: a Reader*, London: Routledge.

Mitchell, Juliet. (1975). *Psychoanalysis and Feminism*, Harmondsworth: Penguin.

Modern Law Review. (2006). Workshop on gender and land rights: University of Keele, Keele, Staffordshire.

Moffett, Helen. (2006). 'These women, they force us to rape them: rape as narrative of social control', *J. of Southern African Studies* 32(1): 129–44.

Mogale, Constance and Poshoko, Sophie. (1997). 'The women's rights in land workshop', *Agenda*, 32: 66–68.

Moghadam, Valentine. (1992). *Privatization and Democratization in Central and Eastern Europe and the Soviet Union*, Wider, United Nations University.

Mohan, Giles; Brown, Ed; Milward, Bob and Zack-Williams, Alfred B. (2000). *Structural Adjustment: theory, practice and impact*, Routledge: London.

Molyneux, Maxine. (1979). 'Beyond the domestic labour debate', *New Left Review*, 116: 3–28.

———. (1985). 'Moblization without emancipation? women's interests, the state and revolution in Nicaragua', *Feminist Studies*: 11(2): 227–54.

———. (1988). 'The politics of abortion in Nicaragua: revolutionary pragmatism, or feminism in the realm of necessity?', *Feminist Review*, 29: 114–32.

———. (2001). *Women's Movements in International Perspective*, London: Institute of Latin American Studies (ILAS) of the University of London.

Momsen, Janet H. (1988). 'Changing gender roles in Caribbean peasant agriculture', in J. Brierly and H. Rubenstein (eds) *Small Farms and Peasant Research in the Caribbean*, Winnipeg: University of Manitoba Dept. of Geography.

———. (2004). *Gender and Development*, London: Routledge.

Momsen, Janet; Szörényi, Irén K. and Timar, Judit. (2005). *Gender at the Border: Enrepreneurship in rural post-socialist Hungary*, Aldershot: Ashgate.

Monbiot, George. (2008). 'These objects of contempt are now our best hope of feeding the world', *Guardian*, 10 June: 25.

Montgomery, J. (1984). *International Dimensions of Land Reform*, Boulder: Westview.

Montoya, Rosario. (2003). 'House, street, collective: revolutionary geographies and gender transformation in Nicaragua, 1979–99', Latin American Research Review, 38(2): 61–93.

Montsalve Suárez, Sofia. (2006). 'Gender and land' in Peter *Rosset*, Raj Patel and Michael Courville *Promised Land: competing visions of agrarian reform*, Oakland: Food First Books: 192–207.

Montsho, Molaole. (2008). 'Zuma says land reform must be speeded up', *Mail and Guardian Online*, 1/5/08. Online. Available HTTP: <http://www.mg.co.za/article/2008–05–01-zuma-says-land-reform-must-be-speeded-up> (accessed 27 July 2008).

Moock, Joyce. (1986). 'Introduction', in *Understanding Africa's Rural Households and Farming Systems*, Boulder: Westview.

Moore, Henrietta. (1988). *Feminism and Anthropology*, Cambridge: Polity.

———. (1994). *A Passion for Difference: essays in anthropology and gender*, Cambridge: Polity.

Moseley, William. (2006). web communication in Cousins, Benjamin. (2006). 'Debating agrarian reform in South Africa: responses to an article in *Mail & Guardian*', (e-mail correspondence with B. Cousins). 8 August 2006'.

Moser, Caroline. (1993). *Gender Planning and Development*. London: Routledge.

Mouzelis, Nicos. (1990). *Post-Marxist Alternatives*, London: Macmillan.

———. (1991). *Back to Sociological Theory: the construction of social orders*, London: Macmillan.

——. (2005). Sectors and Collectives: how the MST Organises its Work'. Available at www.mstbrazil.org/?q=book/print/88, accessed 26 December 2006.

Moyo, Sam. (1995). *The Land Question in Zimbabwe*, Harare: SAPES Trust.

——. (2000). *Land Reform under Structural Adjustment in Zimbabwe*, Uppsala: Nordic Africa Institute.

Moyo, Sam and Yeros, Paris (eds). (2005). *Reclaiming the Land*, London: Zed Books.

Mpundu, Mildred. (2006). 'We know no other home than this: land disputes in Zambia' *Panos Features*. Online. Available HTTP: <http://www.research4development.info/PDF/Outputs/Panos/Zambialand.pdf> (accessed 2 February 2009).

MST (Movimento dos Trabalhadores Rurais Sem Terra) (2004) 'Landless Workers' Movement (MST): MST Social Projects. Online. Available HTTP: <http://www.mstbrazil.org/summary.html> (accessed 29 June 2006).

MST (Movimento dos Trabalhadores Rurais Sem Terra) (2006) 'What is the MST?'. Online. Available HTTP: <http://www.mstbrazil.org/?q=book/print/38> (accessed 24 December 2006).

——. (2009). Movimento dos Trabalhadores Rurais Sem Terra. Online. Available HTTP: <http://www.mst.org.br/mst/home.php> (accessed 2 February 2009).

Muldavin, Joshua. (1998). 'Agrarian change in contemporary rural China', in Ivan Szelényi (ed) *Privatising the Land: political economy in post-communist societies*, London: Routledge: 92–123.

Mulder, Anneke. (1988). 'The Participation of Women in the Agricultural Cooperatives in Maputo, Mozambique', unpublished M.A. Thesis, London School of Economics, University of London.

Muleya, Dumisani. (2000). 'Absence of land funding could be fatal', *Zimbabwe Independent*, 1/9/2000.

Murray, Colin. (1981). *Families Divided: the impact of migrant labour in Lesotho*, Johannesburg: Ravan Press.

Murray, Mary. (1994). *The Law of the Father? Patriarchy in the Transition from Feudalism to Capitalism*, London: Routledge.

Mwagiru, Makumi. (1998). 'Women's land and property rights in three Eastern African countries'. Online. Available HTTP: <www.oxfam.org.uk/http://what_we_do/issues/liveilhoods/landrights/downloads> (accessed 29 August 2002).

Mwaka, Victoria M. (1993). 'Agricultural production and women's time budgets in Uganda', in J. Momsen and V. Kinneard (eds) *Different Places, Different Voices*, London: Routledge.

Myers, Ramon. (1971). *Chinese Peasant Economy: agricultural development in Hopei and Shantung*, Cambridge: Harvard University Press.

Myhre, David. (1996). 'Appropriate agricultural credit: a missing piece of agrarian reform in Mexico', in Laura Randall (ed) *Reforming Mexico's Agrarian Reform*, Armonk, New York/London: M.E. Sharpe: 117–38.

Navarro, Zander. (2005). 'Transforming rights into social practices? the Landless Movement in Brazil', *IDS Bulletin*, 36(1): 129–41.

Nolan, Peter. (1988). *The Political Economy of Collective Farms: an analysis of China's Post-Mao rural reforms*, Polity: Cambridge.

Norfolk, Simon. (2004). 'Examining access to natural resources and linkages to sustainable development', FAO/Food and Agriculture Organization, LSP Livelihoods Support Programme Paper no. 17: Geneva: FAO. Online. Available HTTP: <http://www.fao.org/docref/007/j3619e/13619e00.htm> (accessed 21 December 2006).

Nove, Alex. (1992). *Economic History of the USSR 1917–1991*, Harmondsworth: Penguin.

Ntsebeza, Lungisile. (2005). 'Rural governance and citizenship in post-1994 South Africa: democracy compromised?', in R. Southall, J. Daniel and J. Lutchman (eds) *The State of the Nation 2004–5*, Cape Town: HSRC Press.

———. (2007). 'Address the land question', *Mail and Guardian Online*, 19/8/07. Online. Available HTTP: <http://www.mg.co.za/article/2007–08–19-address-the-land-question> (accessed 2 February 2009).

Ntsebeza, Lungisile and Hall, Ruth. (2007). *The Land Question in South Africa: the challenge of transformation and redistribution*, South Africa: HSRC Press. Online. Available HTTP: <http://www.hsrcpress.ac.za/product. php?productid=2181> (accessed 2 February 2009).

Nuitjen, Monique. (1998). *In the Name of the Land: organization, transnationalism and the culture of the state in a Mexican ejido*, Waginengen: University of Wageningen, PhD thesis.

Oakley, Peter. (1980). 'Participation in development in N.E. Brazil', *Community Development Journal*, 15(1): 10–22.

Odgaard, Rie. (2002). 'Scrambling for land in Tanzania: process of formalisation and legitimisation of land rights', *The European Journal of Development Research*, 14(2): 1–88.

O'Laughlin, Bridget. (2001). 'Gendered poverty, gendered growth' in R. Waterhouse and Carin Vijfhuizen (eds) *Strategic Women, Gainful Men*, Maputo: Universidade Eduardo Mondlane and ActionAid.

Onsrud, Hazel; Paixao, Silvane and Nichols, Sue. (2005). *Women and Land Reform in Brazil*, Dept. of Geodesy and Geomantics Engineering, Technical Report No. 239, Fredericton: University of New Brunswick, Canada.

Pallister, David. (2000). 'Cheap land for absent 'farmers'', *The Guardian*, 28 April. Online. Available HTTP: <http://www.guardian.co.uk/world/2000/apr/28/zimbabwe.davidpallister> (accessed 2 February 2009).

Palmer, Ingrid. (1985). *Women's Roles and Gender Differences in Development: the NEMOW case*, Hartford, Connecticut: Kumarian Press.

Palmer, Robin. (1977). *Land and Racial Domination in Rhodesia*, London: Heinemann.

Palmer, Robin. (1990). 'Land reform in Zimbabwe: 1980–1990', *African Affairs*, 89(355): 163–81

Palmer, Robin. (2000). 'Robert Mugabe and the rules of the game', E-mail (15 November 2000).

Palmer, Robin. (2005). 'Critical reflections on the role of an international NGO seeking to work globally on land rights—with specific focus on Oxfam's experiences in southern Africa'. Online. Available HTTP: www.oxfam.org.uk/.what_we_do/issues/liveilhoods/landrights/downloads> (accessed 10 July 2006).

Pankhurst, Donna. (1991). 'Constraints and incentives in 'successful' Zimbabwean peasant agriculture: the interaction between gender and class', *J. Southern African Studies*, 17(4): 611–32.

Pankhust, Donna and Jacobs, Susie. (1988). 'Land tenure, gender relations and agricultural production: the case of Zimbabwe's peasantry', in J. Davidson (ed) *Women and Land Tenure in Africa*, Boulder: Westview Press.

Pankhurst, Helen. (1992). *Gender, Development and Identity: an Ethiopian study*, London: Zed Books.

Panorama. (2004). *Secrets of the Camps*, 29 February, BBC News.

Pateman, Carol. (1988). *The Sexual Contract*, Stanford: Stanford University Press.

Patiño, Daniel Covarrubias. (1996). 'An opinion survey in the countryside—1994', in Laura Randall (ed) *Reforming Mexico's Agrarian Reform*, Armonk, New York/London: M.E. Sharpe: 107–16.

Patnaik, Utsa. (1979). 'Neo-populism and Marxism: the Chayanovian view of the agrarian question and its fundamental fallacy', *J. Peasant Studies*, 6(4): 375–420.

Paulson, Justin. (2000). 'Peasant struggles and international solidarity: the case of Chiapas', in Leo Panitch and Colin Leys (eds) *The Socialist Register 2000: working classes, global realities*, London/NY: Merlin/Fernwood/Monthly Review Presses.

Pearson, Ruth. (1997). 'Renegotiating the reproduction bargain: gender analysis of economic transition in Cuba in the 1990s', *Development and Change*, 28: 671–705.

Pearson, Ruth and Jackson, Cecile (eds). (1998). *Feminist Visions of Development*, London: Routledge.

Pena, Nuria, Mar Maiques and Gina Castillo. (2008). 'Using rights-based and gender-analysis arguments for land rights for women: some initial reflections from Nicaragua', *Gender and Development*, 16(1): 55–71.

People and the Planet. (2007). 'Urban population trends' from United Nations, World Urbanization Prospects: the 2005 Revision, 29 January. Online. Available HTTP: <http://www.peopleandplanet.net/doc.php?id=1489>, (accessed 4 April 2007).

People's Daily Online. (2004a). 'Gender disparity needs work: opinion' 21 March. Online. Available HTTP: <http://english.peopledaily.com.cn/200403/21/eng20040321_138078.shtml> (accessed 2 February 2009).

People's Daily Online. (2004b). 'Checking imbalance in gender ratio' 26 May. Online. Available HTTP: <http://www.chinadaily.com.cn/english/doc/2004–05/26/content_333951.htm> (accessed 2 February 2009).

People's Daily Online.. (2004c). 'China to balance unbalanced sex ratio' 15 July. Online. Available HTTP: <http://english.peopledaily.com.cn/200407/15/eng20040715_149701.html> (accessed 2 February 2009).

Permanent Mission of the Republic of Zimbabwe. (2000). Statement by C.W.E. Matumbike, Min of National Affairs, Employment Creation and Cooperatives: 'Women 2000: gender equality and development', Permanent Mission of the Rep. of Zimbabwe to the United Nations, New York, 7 June 2000. Online. Available HTTP: <http://un.org/womenwatch/daw/followup/beijing+5stat/statments/zimbabwe7.htm> (accessed 8 July 2008).

Perrotta, Louise. (1998). 'Land reform in the Russian Federation' in S. Bridger and F. Pine (eds) *Surviving Post-Socialism*, London: Routledge: 148–69.

Peters, Beverly and Peters, John. (1998). 'Women and land tenure dynamics in pre-colonial, colonial and post-colonial Zimbabwe' J. of *Public and International Affairs* 9: 183–203.

Petrie, Ragan; Roth, Michael and Mazvimavi, Kizito. (2003). *Seeking Women Land Owners and Ownership in Zimbabwe*, Madison: University of Wisconsin Press.

Pharoah, Robyn. (2001). 'Social Capital and Women's Participation in Three Land Reform Trusts: a case of mixed blessings', unpublished MA thesis, University of Natal, Durban.

Phimister, Ian. (1983). 'Zimbabwe: the path of capitalist development', in D. Birmingham and P. Martin (eds) *A History of Central Africa*, Harlow: Longmans.

Piña-Cabral, João de. (1984). 'Female power and the inequality of wealth and motherhood in northwestern Portugal', in R. Hirschon (ed) *Women and Property—Women as Property*, Beckenham, Kent: Croom Helm.

Pine, Frances. (1986). 'Redefining women's work in rural Poland', in R. Abrahams (ed) *After Socialism*, Oxford: Berghan Books.

———. (1988). 'Dealing with fragmentation: the consequences of privatisation for rural women in central and southern Poland', in S. Bridger and F. Pine (eds) *Surviving Post Socialism*, London: Routledge.

Plaza, Rosío Córdova. (2000). 'Gender roles, inheritance patterns and female access to land in an ejidal community in Veracruz, Mexico', in A. Zoomers and G. van der Haar (eds) *Current Land Policy in Latin America*,Amsterdam: KIT/ Royal Tropical Institute: 161–75.

Pollert, Anna. (1996). 'Gender and class revisited: or, the poverty of 'patriarchy'', *Sociology*, 30(4): 639–60.

Pomfret, John. (2001). 'In China's Countryside, "It's a boy!" too often'. *Washington Post*. Online. Available HTTP: <http://www.puaf.umd.edu/puaf610/ notes/04-Probability_distributions/WP-01–05–29-China-boy-girl-ratio.htm> (accessed 2 February 2009).

Post, Ken. (1972). '"Peasantization' and rural political movements in West Africa', *Archives europeenes de sociologie*, XIII (2): 223–54.

Poston, Dudley and Glover, Karen. (2004). 'Too many males: marriage market implications of gender imbalances in China', Texas A & M University. Online. Available HTTP: <http://iussp2005.princeton.edu/download. aspx?submissionId=50404> (accessed 22 December 2008).

Potter, Jack and Potter, Shulamith. (1990). *China's Peasants: an anthropology of a revolution*, New York: Cambridge University Press.

Powelson, John and Stock, Richard. (1987). *The Peasant Betrayed: Agriculture and Land Reform in the Third World*, Boston: Lincoln Institute/ Oelgeshlager, Gunn and Hain.

Premat, Adriana. (2003). 'Small-scale urban agriculture in Havana and the reproduction of the 'new man' in contemporary Cuba', *Revista Europea de Estudios latinoamericanos y del Caribe*, 75: 85–99.

Prosterman, Roy and Riedinger, Jeffrey M. (1987). *Land Reform and Democratic Development*, Baltimore: Johns Hopkins University Press.

Prosterman, Roy; Temple, Mary and Hanstad, Timothy (eds). (1990). *Agrarian Reform and Grassroots Development: ten case studies*, Boulder: Lynne Rienner.

Pryor, Fredric. (1992). *The Red and the Green: the rise and fall of collectivized agriculture in Marxist regimes*, Princeton: Princeton University Press.

Putzel, James. (1992). *A Captive Land:politics of agrarian reform in the Philippines*, London: Catholic Institute for International Relations.

Raftopoulos, Brian. (2002). 'Key note address: the crisis in Zimbabwe', paper presented at the Canadian Association of African Studies Annual Conference, University of Toronto, Toronto, Ontario, 29 May 2002.

Raghunath, Usha. (1996). 'Land reforms and women agricultural labourers: case studies in Nellore District', in B. N. Yugandhar (ed) *Land Reforms in India*: *Andhra Pradesh*, Vol. 3, London: Sage: 340–66.

Rai, Shirin M. (1995). 'Gender in China', in R. Benewick and P. Wingrove (eds) *China in the 1990s*, Basingstoke: Macmillan: 181–92.

———. (2002). *Gender and the Political Economy of Development*, Cambridge: Polity Press.

Rajhuram, Parvati. (1993). 'Invisible female agricultural labour in India' in J. Momsen and V. Kinneard (eds) *Different Places, Different Voices*, London: Routledge.

Rama, Martin. (2002). 'The gender implications of public sector downsizing: the reform program of Vietnam', *World Bank Research Observer* 17(2): 167–89.

Ramesh, Randeep. (2006). 'Reform or perish: Marxist state's sunrise tinges red rupee pink', *The Guardian*, 5 January. Online. Available HTTP: <http://www. guardian.co.uk/business/2006/jan/05/india.internationalnews> (accessed 9 July 2008).

———. (2007). 'Poor but defiant, thousands march on Delhi in fight for land rights', *The Guardian*, 25 September. Online. Available HTTP: <http://www.guardian.co.uk/world/2007/oct/25/india.randeepramesh> (accessed 2 February 2009).

———. (2008). 'World's cheapest car comes at a high price for West Bengal's farming classes' *The Guardian*, 8 September. Online. Available HTTP: <http://www.guardian.co.uk/world/2008/sep/08/india.automotive> (accessed 2 February 2009).

Ranger, Terence. (1983). 'The invention of tradition in colonial Africa,' in E. Hobsbawm and T. Ranger *The Invention of Tradition*, Cambridge: Cambridge University Press.

———. (1985). *Peasant Consciousness and Guerrilla War in Zimbabwe: a comparative study*, Oxford: James Currey.

Ramazanoglu, Caroline. (1989). *Feminism and the Contradictions of Oppression*, London: Routledge.

Razavi, Shahrashoub. (1994). 'Agrarian change and gender relations in southeast Iran', *Development and Change*, 25(3): 591–634.

———. (ed) (2002). *Shifting Burdens: gender and agrarian change under neoliberalism*, New Haven: Kumarian Press.

———. (2006). 'Land Tenure Reform and Gender Equity', UNRISD Briefing 4.

Redclift, Nanneke and Mignione, E. (eds) (1985) *Beyond Employment: household, gender and subsistence*, Oxford: Blackwell.

Redfield, Robert. (1956). *Peasant Society and Culture*, Chicago: University of Chicago Press.

Répassy, Helga. (1991). 'Changing gender roles in Hungarian agriculture' ,*J. of Rural Studies* 7(1–2): 23–29.

Reuters. (2009). 'Zimbabwe Cholera Deaths more that 2000'. Available at : http://reuters.com/article/worldnews/idUSTRE50C30620090113, accessed 25 January 2009.

Rey, Pierre-Paul. (1975). 'The Lineage Mode of Production', *Critique of Anthropology*, 3: 27–79.

Rice, Xan. (2008). 'Treason threat to party official as Tsvangirai is detained again', *The Guardian*, 13/6/08. Online. Available HTTP: <http://www.guardian.co.uk/world/2008/jun/13/zimbabwe> (accessed 6 July 2008).

Rich, A. (1986). 'Compulsory heterosexuality and lesbian existence' in Adrienne Rich *Blood, Bread and Poetry*, New York: Norton.

Riedinger, Jeffrey M. (1990). 'Philippine land reform in the 1980s', in R. Prosterman, M. Temple and T. Hanstad (eds) *Agrarian Reform and Grassroots Development: ten case studies*, Boulder: Lynne Rienner.

Rigg, Jonathan. (2006). 'Land, farming, livelihoods, and poverty: rethinking the links in the rural South', *World Development* 34(1): 180–202.

Roberts, Pepe. (1983). 'Feminism *in* Africa, feminism *and* Africa', *Review of African Political Economy*, 10(27–28): 175–84.

Rocha, Jan. (2000). *Brazil*, Oxfam Country Profiles, Oxford: Oxfam.

Rogers, Barbara. (1980). *The Domestication of Women: discrimination in developing societies*, London: Tavistock.

———. (1981). 'Land reform—the solution or the problem?', *Human Rights Quarterly*, 3 (2): 96–102.

Rosener, Werner. (1994). *The Peasantry of Europe*, Oxford: Blackwells.

Ross, John. (1996). *Mexico: a guide to the people, politics and culture*, London: Latin America Bureau.

Rosset, Peter. (2006). *Food Is Different: why the WTO should get out of agriculture*, London: Zed Books.

Rothman, Franklin D. and Oliver, Pamela. (1999). 'From local to global: the antidam movement in Southern Brazil: 1979–1992', *Mobilization: an International Journal*, 4 (1): 41–57.

Rowbotham, Sheila. (1982). 'The trouble with patriarchy', in Mary Evans (ed) *The Woman Question*, London: Fontana Books.

Ruben, Ruerd, Rodríguez, Luis and Cortez, Orlando. (2001). 'Land reform, rural organization and agrarian incomes in Nicaragua', in A. Zoomers and G. van der Haar (eds) *Current Land Policy in Latin America*, Amsterdam: KIT/Royal Tropical Institute: 157–68.

Ruben, Ruerd and Masset, Edoardo. (2003). 'Land markets, risk and distress sales in Nicaragua: the impact of income shocks on rural differntiation', *J. of Agrarian Change*: 3(4): October: 481–99.

Rubin, Gayle. (1975). 'The traffic in women', in R. R. Reiter (ed) *Toward an Anthropology of Women*, New York: Monthly Review Press.

Ruchwarger, Gary. (1989). *Struggling for Survival: workers, women and class on a Nicaraguan state farm*, Boulder: Westview.

Rugadya, Margaret, Obaikol, E and Kamusiime, Herbert. (2004). 'Gender and the land reform process in Uganda', Associates for Development. Online. Available HTTP: <http://www.oxfam.org.uk/resources/learning/landrights/downloads/afd_gender_land_reform_process.pdf> (accessed 2 February 2009).

Russell, Alec. (2007). 'Farmers struggle under Mugabe', *Financial Times* (South Africa), 24/5/07.

Sabharwal, G and Thien, Huong T. (2006). 'Missing girls in Vietnam: is high tech sexism an emerging reality?' Online. Available HTTP: <www.eldis.org/cf/rdr/rdr.cfm?doc=DOC16286> (accessed 8 December 2007).

Sachikonye, Lloyd. (2003). 'The Situation of Commercial Farm Workers after Land Reform in Zimbabwe' Report for the Farm Community Trust of Zimbabwe. Online. Available HTTP: <www.gg.rhbnc.ac.uk/simon/Farworkers.pdf> (accessed 30 March 2006).

Sachs, Carolyn. (1995). *Gendered Fields: rural women, agriculture, and environment*, Boulder: Westview Press.

Sachs, Jeffrey. (2004). 'The millennium compact and the end of hunger', in J. Miranowski and C. Scanes (eds) *Perspectives in World Food and Agriculture*, Oxford: Blackwell: 33–40.

Safilios-Rothschild, Constantina. (1988). 'The impact of agrarian reform on men's and women's incomes in rural Honduras', in Dwyer, Daisy and Bruce, Judith (eds) *A Home Divided: women and income in the third world*, Stanford: Stanford University Press.

Sahlins, Marshall. (1974). *Stone Age Economics*, Tavistock: London.

Salvaro, Gionana Ilka J. (2004). 'Jornadas de trabalho de mulheres e homens em um assentamento do MST' *Revista Estudos Feministas*, 12(1): 321–330.

Sanders, David, Cousins, Ben and Moore, David. (2007). 'How not to understand Zimbabwe', *The Mail and Guardian Online*, 28/8/07. Online. Available HTTP: <http://www.mg.co.za/article/2007–08–29-how-not-to-understand-zim> (accessed 2 February 2009).

Scoones, Ian (2008). 'A New Start for Zimbabwe?' Available at www.lalr.org.za/news/a-new-start-for-zimbabwe-by-ian-scoones, accessed 20 January 2009.

Scott, James C. (1985). *Weapons of the weak: everyday forms of peasant resistance*, New Haven, Connecticut: Yale University Press.

Scott, Stephanie. (2003). 'Gender, household headship and entitlements to land: new vulnerabilities in Vietnam's decollectivization', *Gender, Technology and Development* 7(2): 233–263.

Scudder, Thaya. (1973). *The Impact of Integrated River Basin Development on Local Populations*, Vol. I Budapest: Institute for Hydrological Development.

Segal, Lynne. (1999). *Why Feminism?*, Cambridge: Polity Press.

Selden, Mark. (1993). *The Political Economy of Chinese Development*, Armonk, New York: M.E. Sharpe.

——. (1998). 'After collectivization: continuity and change in rural China', in I. Szelényi (ed) *Privatizing the Land*, London: Routledge: 125–=48.

Sen, Amartya. (1981). *Poverty and Famines: an essay on entitlement and deprivation*, Oxford: Clarendon Press.

——. (1990). 'Gender and cooperative conflicts' in I. Tinker (ed) *Persistent Inequalities*, Oxford: Oxford University Press.

Sender, John and Johnston, Deborah. (2004). 'Searching for a weapon of mass production in rural Africa: unconvincing arguments for land reform', *Journal of Agrarian Change*, 4(1–2): 142–64.

Shackleton, Sheona E., Shackleton, Charlie and Cousins, Ben. (2000). 'The economic value of land and natural resources to rural livelihoods: case studies', in B. Cousins (ed) *At the Crossroads: land and agrarian reform in South Africa: into the 21st Century*, Cape Town: PLAAS: 35–67.

Shanin, Teodor. (1974). 'The nature and logic of the peasant economy', *J.Peasant Studies*, 1(1): 63–80.

——. (ed) (1990) *Defining Peasants*, Blackwell: Oxford.

Sihlongonyane, Mfaniseni. (2005). 'Land occupations in South Africa' in S. Moyo and P. Yeros (eds) *Reclaiming the Land*, London and Cape Town: Zed Books.

Sikor, Thomas. (2001). 'Agrarian differentiation in post-socialist societies: evidence from three upland villages in North-Western Vietnam', *Development and Change*, 32: 923–49.

Sikor, Thomas and Pham, T.T. V. (2005). 'Dynamics of commoditization in a Vietnamese uplands village: 1980–2000', *J. Agrarian Change* 5(3): 405–28.

da Silva, Cristiani Bereta. (2004). 'Relações de gênero e subjectividades no devir MST', *Revista Estudos Feministas*, 12(1). Online. Available HTTP: <http://www.scielo.br/scielo.php?script=sci_issuetoc&pid=0104–026X20040001> (accessed 3 December 2006).

de Silva, D. (1982). 'Women and social adaptation in a pioneer Mahweli settlement', in K. Wimalaharma (ed) *Land Settlement Experiences in Sri Lanka*, Colombo: Ministry of Lands: 139–53.

Sobhan, Rehman. (1993). *Agrarian Reform and Social Transformation: preconditions for development*, London: Zed Books.

Solidarity Peace Trust. (2006). 'Operation Taguta/Sisuthi: command Agriculture in Zimbabwe—its impact on rural communities in Matabeleland'. Online. Available HTTP: <http://www.kubatana.net/html/archive/demgg/060405spt.asp?sector=AGRIC> (accessed 2 February 2009).

Song, Hae-Yung. (2009). 'Women and land reform in South Korea'. E-mail (11 January 2009).

Spindel, C. R. (1987). 'The social invisibility of women's work in Brazilian agriculture', in C. D. Deere and M. Leon de Leal (eds) *Rural Women and State Policy: feminist perspectives on agricultural development in Latin America*, Boulder: Westview.

Spring, Anita. (2002). 'Gender and the range of African entrepreneurial strategies: 'typical' and 'new' women entrepreneurs', in A. Jaliloh and T. Falola (eds) *Black Business and Economic Power*, Rochester, NY: University of Rochester Press.

Stacey, Judith. (1982). 'People's war and the new democratic patriarchy in China', *J. Comparative Family Studies*: 13(3): 255–76.

——. (1983). *Patriarchy and Socialist Revolution in China*, Berkeley: University of California Press.

Steele, Jonathan. (1996). 'White collar revolution', *The Guardian*, 12 October. Online. Available HTTP: <http://www.guardian.co.uk/fromthearchive/story/0,,1941553,00.html> (accessed 2 February 2009).

Stephen, Lynn. (1996). 'Too little, too late? The impact of Article 27 on women in Oaxaca', in L. Randall (ed) *Reforming Mexico's Agrarian Reform*, Armonk, New York and London: M.E. Sharpe: 289–303.

———. (1997). *Women and Social Movements in Latin America*, Austin: University of Texas Press.

Stewart, Ann and Tsanga, Amy. (2007). 'The widows' and female child's portion', in S. Ali, A. Hellum, J. Stewart and A. Tsanga (eds) *Human Rights, Plural Legalities and Gendered Realities*, Harare, Zimbabwe: Weaver Press, 407–435.

Stewart, Michael. (1998). "We should build a statue to Ceaucescu here': the trauma of decollectivisation in two Romanian villages', in S. Bridger and F. Pine (eds) *Surviving Post-Socialism*, London: Routledge.

Stivens, Maila; Ng, Cecilia; Jomo K S, Bee, Jahara. (1994). *Malay Peasant Women and the Land*, London: Zed Books.

Stoneman, Colin (ed) (1988) *Zimbabwe's Prospects*, Basingstoke: Macmillan.

Stoneman, Colin. (2000). 'Zimbabwe's land policy and the land reform programme', in T. Bowyer-Bower and C. Stoneman (eds) *Land Reform in Zimbabwe: constraints and prospects*, Aldershot: Ashgate.

Stubbs, Jean. (1987). 'Gender issues in contemporary Cuban tobacco farming', *World Development*, 15(1): 41–65.

———. (1993). 'Women and Cuban smallholder agriculture in transition', in J. Momsen (ed) *Women and Change in the Caribbean: a Pan-Caribbean perspective*, Kingston/Boomington, Indiana/London: I. Randle/Indiana University Press/J. Currey: 219–31.

Stubbs, Jean and Alvarez, Mavis. (1987). 'The cooperative movement in Rural Cuba', in C. D. Deere and M. León de Leal (eds) *Rural Women and State Policy: feminist perspectives on Latin America agricultural development*, Boulder: Westview: 142–61.

Summerfield, Gayle. (2006). *Women and Gender Equity in Development Theory and Practice : institutions, resources, and mobilization*, Chapel Hill: Duke University Press.

Swarns, Rachel. (2002). 'After Zimbabwe's land revolution, new farmers struggle and starve', *The New York Times*, 4/8/08. Online. Available HTTP: <http://query.nytimes.com/gst/fullpage.html?res=9E00E2DC103CF935A15751C1A96 49C8B63> (accessed 5 August 2008).

Swinnen, Johan. (2001). 'Transition from collective farms to individual tenures in Central and Eastern Europe', in A. de Janvry, G. Gordillo, J.-P. Philipe Plateau and E. Sadoulet (eds) *Access to Land, Rural Poverty and Public Action*, Oxford: Oxford University Press: 349–88.

Tadesse, Zenabaworke. (1982). 'The impact of land reform on women: the case of Ethiopia', in L. Beneria (ed) *Women and Development*, New York: Praeger.

———. (2003). 'Women and land rights in the third world: the case of Ethiopia', in L. Muthoni Wanyeki, (ed) *Women and Land in Africa*, London/Cape Town: Zed.

Takavarasha, Tobias. (1998). 'Dimensions of a reformed land structure', *BZS Zimbabwe Review*, 2/98.

Tétreault, Mary Ann. (1994). 'Women and revolution in Vietnam', in M.A. Tétreault (ed) *Women and Revolution in Africa, Asia and the New World*, Charlottesville: University of South Carolina Press.

Therborn, Göran. (2004). *Between Sex and Power: Family in the World 1900–2000*, London: Routledge.

Thiesenhusen, William. (1995). *Broken Promises: Agrarian Reform and the Latin American Campesino*, Boulder: Westview.

———. (1996). 'Mexican land reform 1934–91: success or failure?', in L. Randall (ed) *Reforming Mexico's Agrarian Reform*, Armonk, New York and London: M.E. Sharpe: 35–47.

Thomas, Neil. (2003). 'Land reform in Zimbabwe', *Third World Quarterly*, 24(4): 691–712.

Thorner, Daniel. (1966). 'Introduction' in D. Thorner, B. Kerblay and R. Smith (eds) *The Theory of Peasant Economy*, Homewood, Illinois: R. Irwin.

Thornton A.C. (2009), 'Pastures of plenty? Land rights and community-based agriculture in Peddie, a former homeland town in South Africa', *Applied Geography*, 29(1): 12–20.

Tinker, Irene. (1999). 'Women's empowerment through house and land', in I. Tinker and G. Summerfield (eds) *Women's Rights to House and Land*, Boulder: Lynne Rienner: 9–26.

Tinker, Irene and Summerfield, Gail. (1999). 'Conclusion', in I. Tinker and G. Summerfield (eds) *Women's Rights to House and Land*, Boulder: Lynne Rienner: 265–72.

Tinsman, Heidi. (2002). *Partners in Conflict: the politics of gender, sexuality and labor in the Chilean agrarian reform, 1950–73*, Durham, NC: Duke University Press.

Torres, Rebecca; Momsen, Janet and Niemeier, Debbie. (2007). 'Cuba's farmers' markets in the 'special period' 1990–95', in J. Besson and J. Momsen (eds) *Caribbean Land and Development Revisited*, London: Palgrave.

Toulmin, Camilla and Quan, Julian. (2001). 'Evolving land rights: policy and tenure in sub-Saharan Africa', in C. Toulmin and J. Quan (eds) *Evolving Land Rights: policy and tenure in Africa*, London: DfID/IIED.

Tran Thi van Anh. (1999). 'Women and rural land in Vietnam', in I. Tinker and G. Summerfield (eds) *Women's Rights to House and Land*, Boulder: Lynne Rienner: 95–114.

Tran Thi van Anh and Le Ngoc Hung. (2000). *Women and đổi mới in Vietnam*, Hanoi: Woman Publishing House.

Tripp, Aili Marie. (2004). 'Women's movements, customary law and land rights in Africa: the case of Uganda', *African Studies Quarterly*, 7(4).

———. (2006). 'Challenges in transnational feminist mobilization', in M. M. Ferree and A.M. Tripp (eds) *Transnational Women's Activism, Organizing and Human Rights*, New York: New York University Press: 296–312.

Tsikata, Dzodzai. (2003). 'Securing women's interests within land tenure reforms: recent debates in Tanzania', *J. Agrarian Change* 3(1–2): 149–83.

Tuirán, Rodolfo *et al.* (2005). *Índice de Intensidad Migratoria*, Consejo Nacional de Población, Mexico, D.F.

Turshen, Meredith. (1995). 'Women and health issues', in M. Hay and S. Stichter *African Women South of the Sahara*, Harlow: Longmans, 2nd edition.

———. (2006). 'The role of women in conflict and post-conflict transformation' G-MOSS (Global Monitoring for Security and Stability) [of European Commission] Workshop: *Gender and Security*, 8 March: Lake Maggiore, Italy.

UNICEF. (2006). 'At a glance: Nicaragua, the big picture'. Online. Available HTTP: <http://www.unicef.org/infobycountry/nicaragua.html> (accessed 26 December 2006).

United Nations Development Programme (UNDP) (2001) 'Viet Nam: Gender Briefing Kit', Hanoi: UNDP in Vietnam.

Unterhalter, Elaine. (1987). *Forced Removal*, London: International Defence and Aid Fund For Southern Africa.

Urdang, Stephanie. (1989). *And Still They Dance: women, war and the struggle for change in Mozambique*, New York: Monthly Review Press.

Ushe, Florence. (2007). 'Doomsday predictions premature?' IWPR [Institute for War and Peace Reporting], 21 June 'ZWNews' E-mail (25 June 2007).

Van Schalkwyk, Surika. (2008). 'Food vs Land reform', *Mail and Guardian online*, 12/5/08. Online. Available HTTP: <http://www.mg.co.za/article/2008–05–12-food-vs-land-reform> (accessed 6 July 2008).

Vargas-Lunius, Rosemary with Annelou Ypeij. (2008). *Polishing the Stone: a journey through the promotion of gender equality in development projects*, Rome: IFAD and CEDLA. Online. Available HTTP: <http://www.ifad.org/pub/gender/polishing/polishing.pdf> (accessed 2 February 2009).

Verdery, Katherine. (1990). *What was Socialism, and What Comes Next?* Berkeley: University of California Press.

Viffhuizen, Carin. (1996). 'Who feeds the children?', in Emmanuel Manzungu (ed) *The Practice of Smallholder Irrigation*, Harare, Zimbabwe: University of Zimbabwe Publications, 126–147.

Viola, Lynne. (1996). *Peasant Rebels under Stalin: collectivization and the culture of peasant resistance*, Oxford: Oxford University Press.

Wade, Robert H. (2004). 'Is Globalisation reducing poverty and inequality?' *World Development*, 32(4): 567–89.

Walby, Sylvia. (1990). *Theorising Patriarchy*, Blackwell: Oxford.

Walker, Cherryl. (2003). 'Piety in the sky? Gender policy and land reform in South Africa', *J. Agrarian Change*, 3 (1–2): 113–48.

——. (2005). 'Women, gender policy and land reform in South Africa', *Politikon*, 32(2): 297–315.

——. (2007). '"Redistributive land reform in South Africa: for what and for whom?', in L. Ntsebeza and R. Hall (eds) *The Land Question in South Africa*, Cape Town: HSRC. Online. Available HTTP: <http://www.hsrcpress.ac.za/product.php?productid=2181> (accessed 2 February 2009).

Wang, Lihua. (1999). 'The seeds of socialist ideology: women's experiences in Beishadao village', *Women's Studies International Forum*, 22(1): 25–35.

Wanyeki, Muthoni (ed) (2003) *Women and Land in Africa: culture, religion and realizing women's rights*, London: Zed Books.

Warman, Arturo. (2003). 'Mexico's land reform: a long-term perspective', *Land Reform, Land Settlements and Cooperatives* 2: 84–94.

Washbrook, Sarah. (2005). 'The Chiapas uprising of 1994: historical antecedents and political consequences', *J. of Peasant Studies*: 32(3–4): 417–49.

Waterhouse, Rachel. (1998). 'Women's land rights in post-war Mozambique', in UNIFEM *Women's Land and Property Rights in situations of Conflict and Reconstruction*, Geneva: UNIFEM. Online. Available HTTP: <http://www.oxfam.org.uk/.what_we_do/issues/livelihoods/landrights/downloads/moz-women.rtf> (accessed 20 October 2006).

Waterhouse, Rachel and Vifjhuizen, Carin. (2001). 'Introduction', in R. Waterhouse and C. Vijfhuizen (eds) *Strategic Women, Gainful Men: gender, land and natural resources in different rural contexts in Mozambique*, Maputo: Universidade Eduardo Mondlane and ActionAid.

Waters, Malcolm. (1989). 'Patriarchy and viriarchy', *Sociology* 23(2): 193–211.

Watts, Jonathan. (2005). 'Protests surge as reforms fail to match rising hopes', *Guardian* (London), 11 October: 17.

Watts, Jonathan. (2006). 'Land seizures threaten social stability', *The Guardian*, 21 January. Online. Available HTTP: <http://www.guardian.co.uk/world/2006/jan/21/china.jonathanwatts> (accessed 2 February 2009).

Watts, Michael. (1998). 'Agrarian thermidor: state, decollectivisation and the peasant question in Vietnam', in I. Szelényi (ed) *Privatising the Land*, London: Routledge.

Weber, Max. (2001). *The Protestant Ethic and the Spirit of Capitalism*, London: Routledge (Routledge Modern Classics).

Wegren, Stephen. (1998). 'The conduct and impact of land reform in Russia', in I. Szelényi (ed) *Privatising the Land*, London: Routledge: 3–33.

Wegren, Stephen, O'Brien, David and Patsiorkovski, V. (2002). 'Russian agrarian reform: the gender dimension', *Problems of Post Communism*: 49(6): 48–57.

Wegren, S. K. Patsiorkovski V. V. and O'Brien D. J. (2006). 'Beyond stratification: the emerging class structure in rural Russia', *Journal of Agrarian Change*, 6: 372–99.

Weiner, Daniel. (1988). 'Land and agricultural development', in C. Stoneman (ed) *Zimbabwe's Prospects*, London: Macmillan.

Weinrich, A. K. H. (1975). *African Farmers in Rhodesia*, Oxford: Oxford University Press.

Weissert, Will. (2008). 'Castro gives private farmers room to grow', *Guardian*, 19 July: 8.

Welch, Cliff. (2006). 'Movement histories: a preliminary historiography of the Brazil's Landless Labourers' Movement (MST*)*', *Latin American Research Review*, 41(1): 198–210.

Whatmore, Sarah. (1991). *Farming Women*, London: Macmillan.

White, B. N. (1980). 'Rural household studies in anthropological perspective' in H. Binswanger et al. (eds) *Rural Household Studies in Asia*, Singapore: Singapore University Press.

White, Christine Pelzer. (1982). 'Socialist transformation of agriculture: the Vietnamese case', *Bulletin of the Institute of Development Studies* 13(4).

———. (1987). 'State, culture and gender: continuity and change in women's position in rural Vietnam', in H. Afshar (ed) *Women, State and Ideology in Africa and Asia*, London: Macmillan.

———. (1989). 'Vietnam: war, socialism and the politics of gender', in S. Kruks, R. Rapp and M. Young (eds) *Promissory Notes: women in the transition to socialism*, NY: Monthly Review.

White, Gordon. (1984). 'Developmental states and socialist industrialisation in the third world', *J. Development Studies* 21(1): 97–120.

White, Gordon and Wade, Robert. (1998). 'Developmental states and markets in East Asia: an introduction', in G. White (ed) *Developmental States in East Asia*, Basingstoke: Macmillan: 1–29.

White, Sarah C. (1992). *Arguing with the Crocodile: gender and class in Bangladesh*, London: Zed Books.

Whitehead, Anne. (1981). 'I'm hungry mum—the politics of domestic budgeting', in K. Young, C. Wolkowitz and R. McCullagh (eds) *Of Marriage and the Market*, London: CSE.

———. (2006). 'Some preliminary notes on the subordination of women', *IDS Bulletin* 37 (4): 27–34.

Whitehead, Anne and Tsikata, Dzodzi. (2003). 'Policy discourses on women's land rights in Sub-Saharan Africa: the implications of the re-turn to the customary', *J. Agrarian Change*, 3(1–2): 67–112.

WHO (World Health Organisation) (2006) 'Zimbabwe: health sector needs assessment'. Online. Available HTTP: <http://www.who.int/hac/donorinfo/cap/zimbabwe_cap2008_eng.pdf> (accessed 2 February 2009).

Whyte, Martin King (2000) 'The perils of assessing trends in gender inequality in China', in B. Entwisle and G. Henderson (eds) *Re-Drawing Boundaries: work, household and gender in China*, Berkeley: University of California Press: 157–67.

Wiergsma, Nan. (1988). *Vietnam: peasant land, peasant revolution*, NY: St. Martin's Press.

———. (1991). Peasant patriarchy and the subversion of the collective in Vietnam, *Review of Radical Political Economics*: 23(3–4): 174–97.

Wilk, Richard. (1984). 'Households in process: agricultural change and domestic transformation among the Kekchi Maya of Belize', in R. Netting, R. Wilk and E. J. Arnould (eds) *Households: comparative and historical studies of the domestic group*, Berkley: University of California Press.

———. (1996). *Economies and Cultures: foundations of economic anthropology*. Boulder, Colorado: Westview Press.

Williams, Gavin. (1996). 'Setting the agenda: a critique of the World Bank's rural restructuring programme for South Africa', *J. of Southern African Studies* 22(1): 139–66.

Wittfogel, Karl. (1971). *Oriental Despotism*, New Haven: Yale University Press.

Wittman, Hannah. (2005). 'Agrarian reform under Lula—marching to a historic crossroads'. Online. Available HTTP: <http://www.mstbrazil.org/?q=node/86> (accessed 2 February 2009).

WLSA (Women and Law in Southern Africa). (1994). *Inheritance in Zimbabwe: law, customs and practices*, Harare: WLSA Trust.

———. (1997). *Paradigms of Exclusion: women's access to resources in Zimbabwe*, Harare, Zimbabwe : WLSA Trust.

Wolf, Eric. (1966). *Peasants*, Englewood Cliffs, New Jersey: Prentice-Hall.

———. (1969). *Peasant Wars of the Twentieth Century*, London: Harper and Row.

Wolf, Margery. (1972). *Women and the Family in Rural Taiwan*, Stanford: Stanford University Press.

———. (1975). 'Women and suicide in China', in M. Wolf and R. Witke (eds) *Women in Chinese Society*, Stanford: Stanford University Press.

———. (1985). *Revolution Postponed: women in contemporary China*, Stanford: Stanford University Press.

Wolpe, Harold. (1972). 'Capitalism and cheap labour power in South Africa: from segregation to apartheid', *Economy and Society*, 1(4): 425–56.

———. (1980). 'Introduction'. In *The Articulation of Modes of Production*, ed. H. Wolpe. London: Routledge.

World Bank. (2003). *Land Policies for Growth and Poverty Reduction: a World Bank Policy Research Report*, Oxford: World Bank/Oxford University Press.

———. 'Viet Nam at a glance' 28 September 2007. Online. Available HTTP: <http://devdata.worldbank.org/AAG/vnm_aag.pdf > (accessed 2 February 2009).

World Guide. (2005). *World Guide 2005–06*, Montevideo and Oxford: Instituto del Tercer Mundo/Oxfam Publications.

Wright, Angus and Wolford, Wendy. (2003). *To Inherit the Earth:the landless movement and the struggle for a new Brazil*, Oakland, California: Food First.

Yang, Dali. (1996). *Calamity and Reform in China*, Stanford: Stanford University Press.

Yang, Xiushi. (2000). 'Interconnections among gender, work and migration: evidence from Zhejiang Province', in B. Entwisle and G. Henderson (eds) *Re-Drawing Boundaries: work, household and gender in China*, Berkeley: University of California Press: 197–213.

Young, Kate. (1981). 'Modes of appropriation and the sexual division of labour: a case study from Oaxaca, Mexico', in A. Kuhn and A. M. Wolpe (eds) *Feminism and Materialism*, London: Routledge Kegan Paul.

Yuval-Davis, Nira. (2006). 'Human/women's rights and feminist transversal politics', in M. M. Ferree and A. M. Tripp (eds) *Global Feminism*, New York: New York University Press.

———. (1997). *Gender and Nation*, London: Sage Publishers.

Zaslavskaya, T.I. and Korel, L.V. (1984). 'Rural-urban migration in the USSR: problems and prospects', *Sociologia Ruralis*, 24(3–4): 167–300.

Zhang, Heather. (2003). 'Gender differences in inheritance rights: observations from a Chinese village', *J. Peasant Studies*, 30: 3–4: 252–77.

Zhang, Linxiu. (2001). 'The Situation of Rural Land Contracts: a gender perspective' Workshop presentation on women's land rights, Beijing, May.

Zhang, Weiguo. (2000). 'Dynamics of marriage change in Chinese rural society in transition', *Population Studies*, 5(4): 57–69.

Zhou, Kate. (1996). *How the Farmers Changed China*, Boulder: Lynne Rienner.

Zimyana, Lovemore. (1995). 'The sustainability of smallholder food systems in Africa: the case of Zimbabwe', in T. Binns (ed) *People and Environment in Africa*, Chichester: Wiley Pubs.

Zoomers, Annelies. (2000). 'Land in Latin America: new context, new claims, new concepts', in A. Zoomers and G. van der Haar (eds) *Current Land Policy in Latin America*, Amsterdam: KIT/Royal Tropical Institute: 59–72.

Zoomers, A. and G. van der Haar. (2000). 'Introduction: regulating land policy under neo-liberalism' in Annelies Zoomers and Gemma van der Haar (eds) *Current Land Policy in Latin America*, Amsterdam: KIT/Royal Tropical Institute: 17–28.

Zulu, S. (1999). 'Zimbabwe Court rules women are teenagers', *Mail & Guardian*: 7, 13 May. Online. Available HTTP: <http://www.aegis.org/news/dmg/1999/MG990503.html> (accessed 2 February 2009).

Zuo Jiping. (2004). 'Feminization of agriculture, relational exchange and perceived fairness in China: a case in Guangxi Province', *Rural Sociology*, 69(4): 510–31.

Zweig, David. (1989). 'Struggling over land in China: peasant resistance after collectivization', in F. Colburn (ed) *Everyday Forms of Peasant Resistance*, New York: M. E. Sharpe: 151–74.

ZWRCN (Zimbabwe Women's Resource Centre and Network). (1996). *Woman Plus*, Special Issue: *Women and Land*, Harare: ZWRCN.

Index

Index

T - #0222 - 111024 - C0 - 229/152/13 - PB - 9780415807999 - Gloss Lamination